MEMOIRS
of the
American Mathematical Society
ISSN 0065-9266 (print)
ISSN 1947-6221 (online)

Number 1346

Elliptic Theory for Sets with Higher Co-Dimensional Boundaries

G. David
J. Feneuil
S. Mayboroda

November 2021 • Volume 274 • Number 1346 (fourth of 4 numbers)

AMS AMERICAN MATHEMATICAL SOCIETY

Library of Congress Cataloging-in-Publication Data

Cataloging-in-Publication Data has been applied for by the AMS.
See http://www.loc.gov/publish/cip/.
DOI: https://doi.org/10.1090/memo/1346

Memoirs of the American Mathematical Society

This journal is devoted entirely to research in pure and applied mathematics.

Subscription information. Beginning with the January 2010 issue, *Memoirs* is accessible from www.ams.org/journals. The 2021 subscription begins with volume 269 and consists of six mailings, each containing one or more numbers. Subscription prices for 2021 are as follows: for paper delivery, US$1085 list, US$868 institutional member; for electronic delivery, US$955 list, US$764 institutional member. Upon request, subscribers to paper delivery of this journal are also entitled to receive electronic delivery. If ordering the paper version, add US$20 for delivery within the United States; US$80 for outside the United States. Subscription renewals are subject to late fees. See www.ams.org/help-faq for more journal subscription information. Each number may be ordered separately; *please specify number* when ordering an individual number.

Back number information. For back issues see www.ams.org/backvols.

Subscriptions and orders should be addressed to the American Mathematical Society, P. O. Box 845904, Boston, MA 02284-5904 USA. *All orders must be accompanied by payment.* Other correspondence should be addressed to 201 Charles Street, Providence, RI 02904-2213 USA.

Memoirs of the American Mathematical Society (ISSN 0065-9266 (print); 1947-6221 (online)) is published bimonthly (each volume consisting usually of more than one number) by the American Mathematical Society at 201 Charles Street, Providence, RI 02904-2213 USA. Periodicals postage paid at Providence, RI. Postmaster: Send address changes to Memoirs, American Mathematical Society, 201 Charles Street, Providence, RI 02904-2213 USA.

This publication is indexed in *Mathematical Reviews®*, *Zentralblatt MATH*, *Science Citation Index®*, *Science Citation Index™-Expanded*, *ISI Alerting Services℠*, *SciSearch®*, *Research Alert®*, *CompuMath Citation Index®*, *Current Contents®/Physical, Chemical & Earth Sciences.*
This publication is archived in *Portico* and *CLOCKSS.*
Printed in the United States of America.

∞ The paper used in this book is acid-free and falls within the guidelines established to ensure permanence and durability.
Visit the AMS home page at https://www.ams.org/

10 9 8 7 6 5 4 3 2 1 26 25 24 23 22 21

Contents

Abstract

Many geometric and analytic properties of sets hinge on the properties of elliptic measure, notoriously missing for sets of higher co-dimension. The aim of this manuscript is to develop a version of elliptic theory, associated to a linear PDE, which ultimately yields a notion analogous to that of the harmonic measure, for sets of codimension higher than 1.

To this end, we turn to degenerate elliptic equations. Let $\Gamma \subset \mathbb{R}^n$ be an Ahlfors regular set of dimension $d < n-1$ (not necessarily integer) and $\Omega = \mathbb{R}^n \setminus \Gamma$. Let $L = -\operatorname{div} A\nabla$ be a degenerate elliptic operator with measurable coefficients such that the ellipticity constants of the matrix A are bounded from above and below by a multiple of $\operatorname{dist}(\cdot,\Gamma)^{d+1-n}$. We define weak solutions; prove trace and extension theorems in suitable weighted Sobolev spaces; establish the maximum principle, De Giorgi-Nash-Moser estimates, the Harnack inequality, the Hölder continuity of solutions (inside and at the boundary). We define the Green function and provide the basic set of pointwise and/or L^p estimates for the Green function and for its

---

Received by the editor February 25, 2017, and, in revised form, July 8, 2018.

Article electronically published on November 16, 2021.

DOI: https://doi.org/10.1090/memo/1346

2020 *Mathematics Subject Classification.* Primary 31B05, 35J25; Secondary 28A75, 28A78 , 35J70, 42B20, 42B25, 42B37.

Key words and phrases. Harmonic measure, boundary of co-dimension higher than 1, trace theorem, extension theorem, degenerate elliptic operators, maximum principle, Hölder continuity of solutions, de Giorgi-Nash-Moser estimates, Green functions, comparison principle, homogeneous weighted Sobolev spaces.

The first author is affiliated with the Université Paris-Saclay, CNRS, Laboratoire de Mathématiques d'Orsay, 91405, Orsay, France.

The second author is affiliated with the School of Mathematics, University of Minnesota, Minneapolis, Minnesota 55455.

The third author is affiliated with the School of Mathematics, University of Minnesota, Minneapolis, Minnesota 55455.

This research was supported in part by Fondation Jacques Hadamard and by CNRS. The first author was supported in part by the ANR, programme blanc GEOMETRYA ANR-12-BS01-0014, the European Community Marie Curie grant MANET 607643 and H2020 grant GHAIA 777822, and the Simons Collaborations in MSP grant 601941, GD. The third author was supported by the Alfred P. Sloan Fellowship, the NSF INSPIRE Award DMS 1344235, NSF CAREER Award DMS 1220089, the NSF RAISE-TAQ grant DMS 1839077, and the Simons Foundation grant 563916, SM..

We would like to thank the Department of Mathematics at Université Paris-Sud, the Ecole des Mines, and the Mathematical Sciences Research Institute (NSF grant DMS 1440140) for warm hospitality.

Finally, we would like to thank Alano Ancona for stimulating discussions at the early stages of the project and for sharing with us the results of his work.

gradient. With this at hand, we define harmonic measure associated to L, establish its doubling property, non-degeneracy, change-of-the-pole formulas, and, finally, the comparison principle for local solutions.

In another article to appear, we will prove that when Γ is the graph of a Lipschitz function with small Lipschitz constant, we can find an elliptic operator L for which the harmonic measure given here is absolutely continuous with respect to the d-Hausdorff measure on Γ and vice versa. It thus extends Dahlberg's theorem to some sets of codimension higher than 1.

CHAPTER 1

Introduction

The past few years have witnessed remarkable progress in the study of relations between regularity properties of the harmonic measure ω on the boundary of a domain of $\mathbb{R}^n$ (for instance, its absolute continuity with respect to the Hausdorff measure $\mathcal{H}^{n-1}$) and the regularity of the domain (for instance, rectifiability properties of the boundary). In short, the emerging philosophy is that the rectifiability of the boundary is necessary for the absolute continuity of ω with respect to $\mathcal{H}^{n-1}$, and that rectifiability along with suitable connectedness assumptions is sufficient. Omitting for now precise definitions, let us recall the main results in this regard. The celebrated 1916 theorem of F. & M. Riesz has established the absolute continuity of the harmonic measure for a simply connected domain in the complex plane, with a rectifiable boundary [**RR**]. The quantifiable analogue of this result (the A^∞ property of harmonic measure) was obtained by Lavrent'ev in 1936 [**Lv**] and the local version, pertaining to subsets of a rectifiable curve which is a boundary of a simply connected planar domain, was proved by Bishop and Jones in 1990 [**BJ**]. In the latter work the authors also showed that some connectedness is necessary for the absolute continuity of ω with respect to $\mathcal{H}^{n-1}$, for there exists a planar set with rectifiable boundary for which the harmonic measure is singular with respect to $\mathcal{H}^{n-1}$.

The situation in higher dimensions, $n \geq 3$, is even more complicated. The absolute continuity of ω with respect to $\mathcal{H}^{n-1}$ was first established by Dahlberg on Lipschitz graphs [**Da**] and was then extended to non-tangentially accessible (NTA) domains with Ahlfors regular boundary in [**DJ**], [**Se**], and to more general NTA domains in [**Ba**]. Roughly speaking, the non-tangential accessibility is an assumption of quantifiable connectedness, which requires the presence of interior and exterior corkscrew points, as well as Harnack chains. Ahlfors regularity simply postulates that the measure of intersection with the boundary of every ball of radius r centered at the boundary is proportional to r^{n-1}, i.e., that the boundary is in a certain sense $n-1$ dimensional (we will provide a careful definition below). Similarly to the lower-dimensional case, counterexamples show that some topological restrictions are needed for the absolute continuity of ω with respect to $\mathcal{H}^{n-1}$ [**Wu**], [**Z**]. Much more recently, in [**HM1**], [**HMU**], [**AHMNT**], the authors proved that, in fact, for sets with Ahlfors regular boundaries, under a (weaker) 1-sided NTA assumption, the uniform rectifiability of the boundary is equivalent to the complete set of NTA conditions and hence, is equivalent to the absolute continuity of harmonic measure with respect to the Lebesgue measure. Finally, in 2015 the full converse, "free boundary" result was obtained and established that rectifiability is necessary for the absolute continuity of harmonic measure with respect to $\mathcal{H}^{n-1}$ in any dimension $n \geq 2$ (without any additional topological assumptions) [**AHM3TV**]. It was proved simultaneously that for a complement of an $(n-1)$-Ahlfors regular set the A^∞ property of harmonic measure yields uniform rectifiability of the boundary [**HLMN**]. Shortly after, it was established that in an analogous setting ε-approximability and Carleson measure estimates for bounded harmonic functions are equivalent to uniform rectifiability [**HMM1**], [**GMT**], and that analogous results hold for more general elliptic operators [**HMM2**], [**AGMT**].

The purpose of this work is to start the investigation of similar properties for domains with a lower-dimensional boundary Γ. To the best of our knowledge, the only known approach to elliptic problems on domains with higher co-dimensional boundaries is by means of the p-Laplacian operator and its generalizations [**HKM**], [**LN**]. In [**LN**] the authors worked with an associated Wiener capacity, defined p-harmonic measure, and established boundary Harnack inequalities for Reifenberg flat sets of co-dimension higher than one. Our goals here are different.

We shall systematically assume that Γ is Ahlfors-regular of some dimension $d < n-1$, which *does not need to be an integer.* This means that there is a constant $C_0 \geq 1$ such that

$$C_0^{-1} r^d \leq \mathcal{H}^d(\Gamma \cap B(x,r)) \leq C_0 r^d \quad \text{for } x \in \Gamma \text{ and } r > 0. \tag{1.1}$$

We want to define an analogue of the harmonic measure, that will be defined on Γ and associated to a divergence form operator on $\Omega = \mathbb{R}^n \setminus \Gamma$. We still write the operator as $L = -\operatorname{div} A\nabla$, with $A : \Omega \to \mathbb{M}_n(\mathbb{R})$, and we write the ellipticity condition with a different homogeneity, i.e., we require that for some $C_1 \geq 1$,

$$\operatorname{dist}(x,\Gamma)^{n-d-1} A(x)\xi \cdot \zeta \leq C_1 |\xi|\,|\zeta| \quad \text{for } x \in \Omega \text{ and } \xi, \zeta \in \mathbb{R}^n, \tag{1.2}$$

$$\operatorname{dist}(x,\Gamma)^{n-d-1} A(x)\xi \cdot \xi \geq C_1^{-1} |\xi|^2 \quad \text{for } x \in \Omega \text{ and } \xi \in \mathbb{R}^n. \tag{1.3}$$

The effect of this normalization should be to incite the analogue of the Brownian motion here to get closer to the boundary with the right probability; for instance if $\Gamma = \mathbb{R}^d \subset \mathbb{R}^n$ and $A(x) = \operatorname{dist}(x,\Gamma)^{-n+d+1} I$, it turns out that the effect of L on functions $f(x,t)$ that are radial in the second variable $t \in \mathbb{R}^{n-d}$ is the same as for the Laplacian on $\mathbb{R}^{d+1}_+$. In some sense, we create Brownian travelers which treat Γ as a "black hole": they detect more mass and they are more attracted to Γ than a standard Brownian traveler governed by the Laplacian would be.

The purpose of the present manuscript is to develop, with merely these assumptions, a comprehensive elliptic theory. We solve the Dirichlet problem for $Lu = 0$, prove the maximum principle, the De Giorgi-Nash-Moser estimates and the Harnack inequality for solutions, use this to define a harmonic measure associated to L, show that it is doubling, and prove the comparison principle for positive L-harmonic functions that vanish at the boundary. Let us discuss the details.

We first introduce some notation. Set $\delta(x) = \operatorname{dist}(x,\Gamma)$ and $w(x) = \delta(x)^{-n+d+1}$ for $x \in \Omega = \mathbb{R}^n \setminus \Gamma$, and denote by σ the restriction to Γ of $\mathcal{H}^d$. Denote by $W = \dot{W}^{1,2}_w(\Omega)$ the weighted Sobolev space of functions $u \in L^1_{loc}(\Omega)$ whose distribution gradient in Ω lies in $L^2(\Omega, w)$:

$$W = \dot{W}^{1,2}_w(\Omega) := \{u \in L^1_{loc}(\Omega) : \nabla u \in L^2(\Omega, w)\}, \tag{1.4}$$

and set $\|u\|_W = \left\{ \int_\Omega |\nabla u(x)|^2 w(x) dx \right\}^{1/2}$ for $f \in W$. Finally denote by $\mathcal{M}(\Gamma)$ the set of measurable functions on Γ and then set

$$H = \dot{H}^{1/2}(\Gamma) := \left\{ g \in \mathcal{M}(\Gamma) : \int_\Gamma \int_\Gamma \frac{|g(x) - g(y)|^2}{|x-y|^{d+1}} d\sigma(x) d\sigma(y) < \infty \right\}. \tag{1.5}$$

Before we solve Dirichlet problems we construct two bounded linear operators $T : W \to H$ (a trace operator) and $E : H \to W$ (an extension operator), such that $T \circ E = I_H$. The trace of $u \in W$ is such that for σ-almost every $x \in \Gamma$,

$$Tu(x) = \lim_{r \to 0} \fint_{B(x,r)} u(y)dy := \lim_{r \to 0} \frac{1}{|B(x,r)|} \int u(y)dy, \tag{1.6}$$

and even, analogously to the Lebesgue density property,

$$\lim_{r \to 0} \fint_{B(x,r)} |u(y) - Tu(x)|dy = 0. \tag{1.7}$$

Similarly, we check that if $g \in H$, then

$$\lim_{r \to 0} \fint_{\Gamma \cap B(x,r)} |g(y) - g(x)|d\sigma(y) = 0 \tag{1.8}$$

for σ-almost every $x \in \Gamma$. We typically use the fact that $|u(x) - u(y)| \leq \int_{[x,y]} |\nabla u| d\mathcal{L}^1$ for almost all choices of x and $y \in \Omega$, for which we can use the absolute continuity of $u \in W$ on (almost all) line segments, plus the important fact that, by (1.1), $\Gamma \cap \ell = \emptyset$ for almost every line ℓ.

Note that the latter geometric fact is enabled specifically by the higher co-dimension ($d < n - 1$), even though our boundary can be quite wild. In fact, a stronger property holds in the present setting and gives, in particular, Harnack chains. There exists a constant $C > 0$, that depends only on C_0, n, and $d < n - 1$, such that for $\Lambda \geq 1$ and $x_1, x_2 \in \Omega$ such that $\operatorname{dist}(x_i, \Gamma) \geq r$ and $|x_1 - x_2| \leq \Lambda r$, we can find two points $y_i \in B(x_i, r/2)$ such that $\operatorname{dist}([y_1, y_2], \Gamma) \geq C^{-1} \Lambda^{-d/(n-d-1)} r$. That is, there is a thick tube in Ω that connects the two $B(x_i, r/2)$.

Once we have trace and extension operators, we deduce from the Lax-Milgram theorem that for $g \in H$, there is a unique weak solution $u \in W$ of $Lu = 0$ such that $Tu = g$. For us a weak solution is a function $u \in W$ such that

$$\int_\Omega A(x) \nabla u(x) \cdot \nabla \varphi(x) dx = 0 \tag{1.9}$$

for all $\varphi \in C_0^\infty(\Omega)$, the space of infinitely differentiable functions which are compactly supported in Ω.

Then we follow the Moser iteration scheme to study the weak solutions of $Lu = 0$, as we would do in the standard elliptic case in codimension 1. This leads to the quantitative boundedness (a.k.a. Moser bounds) and the quantitative Hölder continuity (a.k.a. De Giorgi-Nash estimates), in an interior or boundary ball B, of any weak solution of $Lu = 0$ in $2B$ such that $Tu = 0$ on $\Gamma \cap 2B$ when the intersection is non-empty. Precise estimates will be given later in the introduction. The boundary estimates are trickier, because we do not have the conventional "fatness" of the complement of the domain, and it is useful to know beforehand that suitable versions of Poincaré and Sobolev inequalities hold. For instance,

$$\fint_{B(x,r)} |u(y)|dy \leq Cr^{-d} \int_{B(x,r)} |\nabla u(y)| w(y) dy \tag{1.10}$$

for $u \in W$, $x \in \Gamma$, and $r > 0$ such that $Tu = 0$ on $\Gamma \cap B(x,r)$ and, if $m(B(x,r))$ denotes $\int_{B(x,r)} w(y)dy$,

$$\begin{aligned}\Big\{\frac{1}{m(B(x,r))}\int_{B(x,r)}\Big|u(y) - \fint_{B(x,r)} u\Big|^p w(y)dy\Big\}^{1/p} \\ \le Cr\Big\{\frac{1}{m(B(x,r))}\int_{B(x,r)} |\nabla u(y)|^2 w(y)dy\Big\}^{1/2}\end{aligned} \tag{1.11}$$

for $u \in W$, $x \in \overline{\Omega} = \mathbb{R}^n$, $r > 0$, and $p \in \left[1, \frac{2n}{n-2}\right]$ (if $n \ge 3$) or $p \in [1, +\infty)$ (if $n = 2$).

A substantial portion of the proofs lies in the analysis of the newly defined Sobolev spaces. It is important to note, in particular, that we prove the density of smooth functions on $\mathbb{R}^n$ (and not just Ω) in our weighted Sobolev space W. That is, for any function f in W, there exists a sequence $(f_k)_{k\ge1}$ of functions in $C^\infty(\mathbb{R}^n) \cap W$ such that $\|f - f_k\|_W$ tends to 0 and f_k converges to f in $L^1_{loc}(\mathbb{R}^n)$. In codimension 1, this sort of property, just like (1.10) or (1.11), typically requires a fairly nice boundary, e.g., Lipschitz, and it is quite remarkable that here they all hold in the complement of any Ahlfors-regular set. This is, of course, a fortunate outcome of working with lower dimensional boundary: we can guarantee ample access to the boundary (cf., e.g., the Harnack "tubes" discussed above), which turns out to be sufficient despite the absence of traditionally required "massive complement". Or rather one could say that the boundary itself is sufficiently "massive" from the PDE point of view, due to our carefully chosen equation and corresponding function spaces.

With all these ingredients, we can follow the standard proofs for elliptic divergence form operators. When u is a solution to $Lu = 0$ in a ball $2B \subset \Omega$, the De Giorgi-Nash-Moser estimates and the Harnack inequality in the ball B don't depend on the properties of the boundary Γ and thus can be proven as in the case of codimension 1. When $B \subset \mathbb{R}^n$ is a ball centered on Γ and u is a weak solution to $Lu = 0$ in $2B$ whose trace satisfies $Tu = 0$ on $\Gamma \cap 2B$, the quantitative boundedness and the quantitative Hölder continuity of the solution u are expressed with the help of the weight w. There holds, if $m(2B) = \int_{2B} w(y)dy$,

$$\sup_B u \le C\left(\frac{1}{m(2B)}\int_{2B} |u(y)|^2 w(y)dy\right)^{1/2} \tag{1.12}$$

and, for any $\theta \in (0, 1]$,

$$\sup_{\theta B} u \le C\theta^\alpha \sup_B u \le C\theta^\alpha \left(\frac{1}{m(2B)}\int_{2B} |u(y)|^2 w(y)dy\right)^{1/2}, \tag{1.13}$$

where θB denotes the ball with same center as B but whose radius is multiplied by θ, and $C, \alpha > 0$ are constants that depend only on the dimensions d and n, the Ahlfors constant C_0 and the ellipticity constant C_1.

We establish then the existence and uniqueness of a Green function g, which is roughly speaking a positive function on $\Omega \times \Omega$ such that, for all $y \in \Omega$, the function $g(., y)$ solves $Lg(., y) = \delta(y)$ and $Tg(., y) = 0$. In particular, the following pointwise estimates are shown:

$$(1.14)\qquad 0 \leq g(x,y) \leq \begin{cases} C|x-y|^{1-d} & \text{if } 4|x-y| \geq \delta(y) \\ \frac{C|x-y|^{2-n}}{w(y)} & \text{if } 2|x-y| \leq \delta(y),\ n \geq 3 \\ \frac{C_\epsilon}{w(y)} \left(\frac{\delta(y)}{|x-y|}\right)^\epsilon & \text{if } 2|x-y| \leq \delta(y),\ n = 2, \end{cases}$$

where $C > 0$ depends on d, n, C_0, C_1 and $C_\epsilon > 0$ depends on d, C_0, C_1, ϵ. When $n \geq 3$, the pointwise estimates can be gathered to a single one, and may look more natural for the reader: if $m(B) = \int_B w(y)dy$,

$$(1.15)\qquad 0 \leq g(x,y) \leq C \frac{|x-y|^2}{m(B(x,|x-y|))}$$

whenever $x, y \in \Omega$. The bound in the case where $n = 2$ and $2|x-y| \leq \delta(y)$ can surely be improved into a logarithm bound, but the bound given here is sufficient for our purposes. Also, our results hold for any d and any n such that $d < n-1$, (i.e., even in the cases where $n = 2$ or $d \leq 1$), which proves that Ahlfors regular domains are 'Greenian sets' in our adapted elliptic theory. Note that contrary to the codimension 1 case, the notion of the fundamental solution in $\mathbb{R}^n$ is not accessible, since the distance to the boundary of Ω is an integral part of the definition of L.

We use the Harnack inequality, the De Giorgi-Nash-Moser estimates, as well as a suitable version of the maximum principle, to solve the Dirichlet problem for continuous functions with compact support on Γ, and then to define harmonic measures ω^x for $x \in \Omega$ (so that $\int_\Gamma g d\omega^x$ is the value at x of the solution of the Dirichlet problem for g). Note that we do not need an analogue of the Wiener criterion (which normally guarantees that solutions with continuous data are continuous up to the boundary and allows one to define the harmonic measure), as we have already proved a stronger property, that solutions are Hölder continuous up to the boundary. Then, following the ideas of [**Ken**, Section 1.3], we prove the following properties on the harmonic measure ω^x. First, the non-degeneracy of the harmonic measure states that if B is a ball centered on Γ,

$$(1.16)\qquad \omega^x(B \cap \Gamma) \geq C^{-1}$$

whenever $x \in \Omega \cap \frac{1}{2}B$ and

$$(1.17)\qquad \omega^x(\Gamma \setminus B) \geq C^{-1}$$

whenever $x \in \Omega \setminus 2B$, the constant $C > 0$ depending as previously on d, n, C_0 and C_1. Next, let us recall that any boundary ball has a corkscrew point, that is for any ball $B = B(x_0, r) \subset \mathbb{R}^n$ centered on Γ, there exists $\Delta_B \in B$ such that $\delta(\Delta_B)$ is bigger than ϵr, where $\epsilon > 0$ depends only on d, n and C_0. With this definition in mind, we compare the harmonic measure with the Green function: for any ball B of radius r centered on Γ,

$$(1.18)\qquad C^{-1} r^{1-d} g(x, \Delta_B) \leq \omega^x(B \cap \Gamma) \leq C r^{1-d} g(x, \Delta_B)$$

for any $x \in \Omega \setminus 2B$ and

$$(1.19)\qquad C^{-1} r^{1-d} g(x, \Delta_B) \leq \omega^x(\Gamma \setminus B) \leq C r^{1-d} g(x, \Delta_B)$$

for any $x \in \Omega \cap \frac{1}{2}B$ which is far enough from Δ_B, say $|x - \Delta_B| \geq \epsilon r/2$, where ϵ is the constant used to define Δ_B. The constant $C > 0$ in (1.18) and (1.19) depends again only on d, n, C_0 and C_1. The estimates (1.18) and (1.19) can be seen as weak versions of the 'comparison principle', which deal only with the Green functions and the harmonic measure and which can be proven by using the specific properties of the latter objects. The inequalities (1.18) and (1.19) are essential for the proofs of the next three results.

The first one is the doubling property of the harmonic measure, which guarantees that, if B is a ball centered on Γ, $\omega^x(2B \cap \Gamma) \leq C\omega^x(B \cap \Gamma)$ whenever $x \in \Omega \setminus 4B$. It has an interesting counterpart: $\omega^x(\Gamma \setminus B) \leq C\omega^x(\Gamma \setminus 2B)$ whenever $x \in \Omega \cap \frac{1}{2}B$.

The second one is the change-of-the-pole estimates, which can be stated as

$$C^{-1}\omega^{\Delta_B}(E) \leq \frac{\omega^x(E)}{\omega^x(\Gamma \cap B)} \leq C\omega^{\Delta_B}(E) \tag{1.20}$$

when B is a ball centered on Γ, $E \subset B \cap \Gamma$ is a Borel set, and $x \in \Omega \setminus 2B$.

The last result is the comparison principle, that says that if u and v are positive weak solutions of $Lu = Lv = 0$ such that $Tu = Tv = 0$ on $2B \cap \Gamma$, where B is a ball centered on Γ, then u and v are comparable in B, i.e.,

$$\sup_{z \in B \setminus \Gamma} \frac{u(z)}{v(z)} \leq C \inf_{z \in B \setminus \Gamma} \frac{u(z)}{v(z)}. \tag{1.21}$$

In each case, i.e., for the doubling property of the harmonic measure, the change of pole, or the comparison principle, the constant $C > 0$ depends only on d, n, C_0 and C_1.

It is difficult to survey a history of the subject that is so classical (in the co-dimension one case). In that setting, that is, in co-dimension one and reasonably nice geometry, e.g., of Lipschitz domains, the results have largely become folklore and we often follow the exposition in standard texts [**GT**], [**HL**], [**Maz**], [**MZ**], [**Sta2**], [**GW**], [**CFMS**]. The general order of development is inspired by [**Ken**]. Furthermore, let us point out that while the invention of a harmonic measure which serves the higher co-dimensional boundaries, which is associated to a linear PDE, and which is absolutely continuous with respect to the Lebesgue measure on reasonably nice sets, is the main focal point of our work, various versions of degenerate elliptic operators and weighted Sobolev spaces have of course appeared in the literature over the years. Some versions of some of the results listed above or similar ones can be found, e.g., in [**A**], [**FKS**], [**Haj**], [**HaK2**], [**HKM**], [**Kil**], [**JW**]. However, the presentation here is fully self-contained, and since we did not rely on previous work, we hope to be forgiven for not providing a detailed review of the corresponding literature. Also, the context of the present paper often makes it possible to have much simpler proofs than a more general setting of not necessarily Ahlfors regular sets. It is perhaps worth pointing out that we work with *homogeneous* Sobolev spaces. Unfortunately, those are much less popular in the literature that their non-homogeneous counterparts, while they are more suitable for PDEs on unbounded domains.

As outlined in [**DFM**], , we intend in subsequent publications to take stronger assumptions, both on the geometry of Γ and the choice of L, and prove that the harmonic measure defined here is absolutely continuous with respect to $\mathcal{H}^d_{|\Gamma}$. For instance, we will assume that d is an integer and Γ is the graph of a Lipschitz function $F : \mathbb{R}^d \to \mathbb{R}^{n-d}$, with a small enough Lipschitz constant. As for A, we will assume that $A(x) = D(x)^{-n+d+1}I$ for $x \in \Omega$, with

$$D(x) = \left\{ \int_\Gamma |x-y|^{-d-\alpha} d\mathcal{H}^d(y) \right\}^{-1/\alpha} \tag{1.22}$$

for some constant $\alpha > 0$. Notice that because of (1.1), $D(x)$ is equivalent to $\delta(x)$; when $d = 1$ we can also take $A(x) = \delta(x)^{-n+d+1}I$, but when $d \geq 2$ we do not know whether $\delta(x)$ is smooth enough to work. In (1.22), we could also replace $\mathcal{H}^d$ with another Ahlfors-regular measure on Γ.

With these additional assumptions we will prove that the harmonic measure described above is absolutely continuous with respect to $\mathcal{H}^d_{|\Gamma}$, with a density which is a Muckenhoupt A_∞ weight. In other words, we shall establish an analogue of Dahlberg's result [**Da**] for domains with a higher co-dimensional boundary given by a Lipschitz graph with a small Lipschitz constant. It is not so clear what is the right condition for this in terms of A, but the authors still hope that a good condition on Γ is its uniform rectifiability. Notice that in remarkable contrast with the case of codimension 1, we do not state an additional quantitative connectedness condition on Ω, such as the Harnack chain condition in codimension 1; this is because such conditions are automatically satisfied when Γ is Ahlfors-regular with a large codimension.

The present paper is aimed at giving a fairly pleasant general framework for studying a version of the harmonic measure in the context of Ahlfors-regular sets Γ of codimension larger than 1, but it will probably be interesting and hard to understand well the relations between the geometry of Γ, the regularity properties of A (which has to be linked to Γ through the distance function), and the regularity properties of the associated harmonic measure.

CHAPTER 2

The Harnack Chain Condition and the Doubling Property

We keep the same notation as in Chapter 1, concerning $\Gamma \subset \mathbb{R}^n$, a closed set that satisfies (1.1) for some $d < n-1$, $\Omega = \mathbb{R}^n \setminus \Gamma$, then $\sigma = \mathcal{H}^d_{|\Gamma}$, $\delta(z) = \operatorname{dist}(z, \Gamma)$, and the weight $w(z) = \delta(z)^{d+1-n}$.

Let us add the notion of measure. The measure m is defined on (Lebesgue-)measurable subset of $\mathbb{R}^n$ by $m(E) = \int_E w(z)dz$. We may write $dm(z)$ for $w(z)dz$. Since $0 < w < +\infty$ a.e. in $\mathbb{R}^n$, m and the Lebesgue measure are mutually absolutely continuous, that is they have the same zero sets. Thus there is no need to specify the measure when using the expressions *almost everywhere* and *almost every*, both abbreviated *a.e.*; conversely the expression *a.e.* without any measure refers to *Lebesgue almost everywhere* or equivalently *m-almost everywhere*.

In the sequel of the article, C will denote a real number (usually big) that can vary from one line to another. The parameters which the constant C depends on are either obvious from context or recalled. Besides, the notation $A \approx B$ will be used to replace $C^{-1}A \leq B \leq CA$.

This section is devoted to the proof of the very first geometric properties on the space Ω and the weight w. We will prove in particular that m is a doubling measure and Ω satisfies the Harnack chain condition.

First, let us prove the Harnack chain condition we stated in Section 1.

LEMMA 2.1. *Let Γ be a d-Ahlfors regular set in $\mathbb{R}^n$, $d < n-1$, that is, assume that* (1.1) *is satisfied. Then there exists a constant $c > 0$, that depends only on C_0, n, and $d < n-1$, such that for $\Lambda \geq 1$ and $x_1, x_2 \in \Omega$ such that* $\operatorname{dist}(x_i, \Gamma) \geq r$ *and $|x_1 - x_2| \leq \Lambda r$, we can find two points $y_i \in B(x_i, r/3)$ such that* $\operatorname{dist}([y_1, y_2], \Gamma) \geq c\Lambda^{-d/(n-d-1)}r$. *That is, there is a thick tube in Ω that connects the two $B(x_i, r/3)$.*

PROOF. Indeed, suppose $x_2 \neq x_1$, set $\ell = [x_1, x_2]$, and denote by P the vector hyperplane with a direction orthogonal to $x_2 - x_1$. Let $\varepsilon \in (0,1)$ be small, to be chosen soon. We can find $N \geq C^{-1}\varepsilon^{1-n}$ points $z_j \in P \cap B(0, r/3)$, such that $|z_j - z_k| \geq 4\varepsilon r$ for $j \neq k$. Set $\ell_j = z_j + \ell$, and suppose that $\operatorname{dist}(\ell_j, \Gamma) \leq \varepsilon r$ for all j. Then we can find points $w_j \in \Gamma$ such that $\operatorname{dist}(w_j, \ell_j) \leq \varepsilon r$. Notice that the balls $B_j = B(w_j, \varepsilon r)$ are disjoint because $\operatorname{dist}(\ell_j, \ell_k) \geq 4\varepsilon r$, and by (1.1)

$$NC_0^{-1}(\varepsilon r)^d \leq \sum_j \sigma(B_j) = \sigma\Big(\bigcup_j B_j\Big) \leq \sigma(B(w, 2r + |x_2 - x_1|)) \leq C_0(2+\Lambda)^d r^d$$

where w is any of the w_j. Thus $\varepsilon^{1-n}\varepsilon^d \leq CC_0^2\Lambda^d$ (recall that $\Lambda \geq 1$), a contradiction if we take $\varepsilon \leq c\Lambda^{-d/(n-d-1)}$, where $c > 0$ depends on C_0 too. Thus we can find j such that $\operatorname{dist}(\ell_j, \Gamma) \geq \varepsilon r$, and the desired conclusion holds with $y_i = x_i + z_j$. $\square$

LEMMA 2.2. *Let $\Gamma \subset \mathbb{R}^n$ be an Ahlfors regular set of dimension d, $d < n-1$. Then for any $\Lambda \geq 1$, we can find a large integer N_Λ and a small real ϵ_Λ such that for any couple of points $x_1, x_2 \in \Omega$ satisfying $\operatorname{dist}(x_i, \Gamma) \geq r$ and $|x_1 - x_2| \leq \Lambda r$, there exists a sequence of balls $(B_i)_{0 \leq i \leq N_\Lambda}$ - called Harnack chain - verifying:*

(i) *B_0 is centered on x_1 and B_{N_Λ} is centered on x_2,*
(ii) *for $0 \leq i \leq N_\Lambda$, $3B_i \subset \Omega$ and B_i has radius at least $\epsilon_\Lambda r$,*
(iii) *for $1 \leq i \leq N_\Lambda$, $B_{i-1} \cap B_i \neq \emptyset$.*

PROOF. This lemma is a corollary of Lemma 2.1. First, in the proof, c shall be the constant given by Lemma 2.1.

Let $\Lambda \geq 1$ and x_1, x_2 as in the lemma. Let y_1, y_2 be the associated points given by Lemma 2.1, which thus verify $y_1 \in B(x_1, r/3)$, $y_2 \in B(x_2, r/3)$, and $\operatorname{dist}([y_1, y_2], \Gamma) \geq c\Lambda^{-d/(n-d-1)} r$.

From there, we define ϵ_Λ as $c\Lambda^{-d/(n-d-1)}/3$ and N_Λ as the smallest integer bigger than $(\Lambda + 2)r/\epsilon_\Lambda + 2$. The sequence $(B_i)_{0 \leq i \leq N_\Lambda}$ is then constructed as follows. First, B_0 is $B(x_1, r/3)$. Then $(B_i)_{1 \leq i < N_\Lambda - 1}$ are balls centered on the segment $[y_1, y_2]$ and of radius $\epsilon_\Lambda r$ - precisely B_i is centered on $[(N_\Lambda - i - 1)y_1 + (i-1)y_2]/[N_\Lambda - 2]$ - and finally B_{N_Λ} is $B(x_2, r/3)$. □

Then, we give estimates on the weight w.

LEMMA 2.3. *There exists $C > 0$ such that*

(i) *for any $x \in \mathbb{R}^n$ and any $r > 0$ satisfying $\delta(x) \geq 2r$,*

$$C^{-1} r^n w(x) \leq m(B(x,r)) = \int_{B(x,r)} w(z) dz \leq C r^n w(x), \tag{2.1}$$

(ii) *for any $x \in \mathbb{R}^n$ and any $r > 0$ satisfying $\delta(x) \leq 2r$,*

$$C^{-1} r^{d+1} \leq m(B(x,r)) = \int_{B(x,r)} w(z) dz \leq C r^{d+1}. \tag{2.2}$$

REMARK 2.4. In the above lemma, the estimates are different if $\delta(x)$ is bigger or smaller than $2r$. Yet the critical ratio $\frac{\delta(x)}{r} = 2$ is not relevant: for any $\alpha > 0$, we can show as well that (2.1) holds whenever $\delta(x) \geq \alpha r$ and (2.2) holds if $\delta(x) \leq \alpha r$, with a constant C that depends on α.

Indeed, we can replace 2 by α if we can prove that for any $K > 1$ there exists $C > 0$ such that for any $x \in \mathbb{R}^n$ and $r > 0$ satisfying

$$K^{-1} r \leq \delta(x) \leq K r \tag{2.3}$$

we have

$$C^{-1} r^{d+1} \leq r^n w(x) \leq C r^{d+1}. \tag{2.4}$$

However, since $w(x) = \delta(x)^{d+1-n}$, (2.3) implies $w(x) \approx r^{d+1-n}$ which in turn gives (2.4).

PROOF. First suppose that $\delta(x) \geq 2r$. Then for any $z \in B(x,r)$, $\frac{1}{2}\delta(x) \leq \delta(z) \leq \frac{3}{2}\delta(x)$ and hence $C^{-1} w(x) \leq w(z) \leq C w(x)$; (2.1) follows.

The lower bound in (2.2) is also fairly easy, just note that when $\delta(x) \leq 2r$, $\delta(z) \leq 3r$ for any $z \in B(x,r)$ and hence

$$m(B(x,r)) \geq \int_{B(x,r)} (3r)^{1+d-n} dz \geq C^{-1} r^{d+1}. \tag{2.5}$$

Finally we check the upper bound in (2.2). We claim that for any $y \in \Gamma$ and any $r > 0$,

$$m(B(y,r)) = \int_{B(y,r)} \delta(\xi)^{d+1-n} d\xi \leq Cr^{d+1}. \tag{2.6}$$

From the claim, let us prove the upper bound in (2.2). Let $x \in \mathbb{R}^n$ and $r > 0$ be such that $\delta(x) \leq 2r$. Thus there exists $y \in \Gamma$ such that $B(x,r) \subset B(y,3r)$ and thanks to (2.6)

$$m(B(x,r)) \leq \int_{B(y,3r)} w(z)dz \leq C(3r)^{d+1} \leq Cr^{d+1}, \tag{2.7}$$

which gives the upper bound in (2.2).

Let us now prove the claim. By translation invariance, we can choose $y = 0 \in \Gamma$. Note that $\delta(\xi) \leq r$ in the domain of integration. Let us evaluate the measure of the set $Z_k = \{\xi \in B(0,r)\,;\, 2^{-k-1}r < \delta(\xi) \leq 2^{-k}r\}$. We use (1.1) to cover $\Gamma \cap B(0,2r)$ with less than $C2^{kd}$ balls B_j of radius $2^{-k}r$ centered on Γ; then Z_k is contained in the union of the $3B_j$, so $|Z_k| \leq C2^{kd}(2^{-k}r)^n$ and $\int_{\xi \in Z_k} \delta(\xi)^{1+d-n} d\xi \leq C2^{kd}(2^{-k}r)^n(2^{-k}r)^{d+1-n} = C2^{-k}r^{d+1}$. We sum over $k \geq 0$ and get (2.6). □

A consequence of Lemma 2.3 is that m is a doubling measure, that is for any $x \in \mathbb{R}^n$ and $r > 0$, one has $m(B(x,2r)) \leq Cm(B(x,r))$ for a constant $C > 0$ independent of x and r. Actually, we can prove the following stronger fact: for any $x \in \mathbb{R}^n$ and any $r > s > 0$, there holds

$$C^{-1}\left(\frac{r}{s}\right)^{d+1} \leq \frac{m(B(x,r))}{m(B(x,s))} \leq C\left(\frac{r}{s}\right)^n. \tag{2.8}$$

Three cases may happen. First, $\delta(x) \geq 2r \geq 2s$ and then with (2.1),

$$\frac{m(B(x,r))}{m(B(x,s))} \approx \frac{r^n w(x)}{s^n w(x)} = \left(\frac{r}{s}\right)^n. \tag{2.9}$$

Second, $\delta(x) \leq 2s \leq 2r$. In this case, note that (2.2) implies

$$\frac{m(B(x,r))}{m(B(x,s))} \approx \frac{r^{d+1}}{s^{d+1}} = \left(\frac{r}{s}\right)^{d+1}. \tag{2.10}$$

At last, $2s \leq \delta(x) \leq 2r$. Note that (2.1) and (2.2) yield

$$\frac{m(B(x,r))}{m(B(x,s))} \approx \frac{r^{d+1}}{s^n w(x)}. \tag{2.11}$$

Yet, $2s \leq \delta(x) \leq 2r$ implies $C^{-1}r^{d+1-n} \leq w(x) \leq Cs^{d+1-n}$ and thus

$$C^{-1}\left(\frac{r}{s}\right)^{d+1} \leq \frac{m(B(x,r))}{m(B(x,s))} \leq C\left(\frac{r}{s}\right)^n. \tag{2.12}$$

which finishes the proof of (2.8).

One can see that the coefficients $d+1$ and n are optimal in (2.8). The fact that the volume of a ball with radius r is not equivalent to r^α for some $\alpha > 0$ will cause some difficulties. For instance, regardless of the choice of p, we cannot have a Sobolev embedding $W \hookrightarrow L^p$ and we have to settle for the Sobolev-Poincaré inequality (1.11).

Another consequence of Lemma 2.3 is that for any $x \subset \mathbb{R}^n$, any $r > 0$, and any nonnegative function $g \in L^1_{loc}(\mathbb{R}^n)$,

$$\frac{1}{|B(x,r)|} \int_{B(x,r)} g(z)dz \leq C \frac{1}{m(B(x,r))} \int_B g(z)w(z)dz. \tag{2.13}$$

Indeed, the inequality (2.13) holds if we can prove that

$$\frac{m(B(x,r))}{|B(x,r)|} \leq Cw(z) \qquad \forall z \in B(x,r). \tag{2.14}$$

This latter fact can be proven as follows:

$$\frac{m(B(x,r))}{|B(x,r)|} \leq \frac{m(B(z,2r))}{|B(x,r)|} \leq Cr^{-n} m(B(z,2r)) \tag{2.15}$$

If $\delta(z) \geq 4r$, then Lemma 2.3 gives $r^{-n}m(B(z,2r)) \leq Cw(z)$. If $\delta(z) \leq 4r$, then $w(z) \geq C^{-1}r^{d+1-n}$ and Lemma 2.3 entails $r^{-n}m(B(z,2r)) \leq Cr^{d+1-n} \leq Cw(z)$. In both cases, we obtain (2.14) and thus (2.13).

We end the section with a corollary of Lemma 2.3.

LEMMA 2.5. *The weight w is in the $\mathcal{A}_2$-Muckenhoupt class, i.e. there exists $C > 0$ such that for any ball $B \subset \mathbb{R}^n$,*

$$\fint_B w(z)dz \fint_B w^{-1}(z)dz \leq C. \tag{2.16}$$

PROOF. Let $B := B(x,r)$. If $\delta(x) \geq 2r$, then for any $z \in B(x,r)$, $C^{-1}w(x) \leq w(z) \leq Cw(x)$ and thus $\fint_B w(z)dz \cdot \fint_B w^{-1}(z)dz \leq Cw(x)w^{-1}(x) = C$. If $\delta(x) \leq 2r$, then (2.2) implies that $\fint_B w(z)dz \leq Cr^{-n}r^{d+1} = Cr^{d+1-n}$. Besides, for any $z \in B(x,r)$, $\delta(z) \leq 3r$ and hence $w^{-1}(z) \leq Cr^{n-d-1}$. It follows that if $\delta(x) \leq 2r$, $\fint_B w(z)dz \cdot \fint_B w^{-1}(z)dz \leq C$. The assertion (2.16) follows. □

CHAPTER 3

Traces

The weighted Sobolev space $W = \dot{W}_w^{1,2}(\Omega)$ and $H = \dot{H}^{1/2}(\Gamma)$ are defined as in Section 1 (see (1.4), (1.5)). Let us give a precison. Any $u \in W$ has a distributional derivative in Ω that belongs to $L^2(\Omega, w)$, that is there exists a vector valued function $v \in L^2(\Omega, w)$ such that for any $\varphi \in C_0^\infty(\Omega, \mathbb{R}^n)$

$$\int_\Omega v \cdot \varphi \, dx = -\int_\Omega u \operatorname{div} \varphi \, dx. \tag{3.1}$$

This definition make sense since $v \in L^2(\Omega, w) \subset L^1_{loc}(\Omega)$. For the proof of the latter inclusion, use for instance Cauchy-Schwarz inequality and (2.13).

The aim of the section is to state and prove a trace theorem. But for the moment, let us keep discussing about the space W. We say that u is absolutely continuous on lines in Ω if there exists $\bar{u}$ which coincides with u a.e. such that for almost every line ℓ (for the usual invariant measure on the Grassman manifold, but we can also say, given any choice of direction v and and a vector hyperplane plane P transverse to v, for the line $x + \mathbb{R}v$ for almost every $x \in P$), we have the following properties. First, the restriction of $\bar{u}$ to $\ell \cap \Omega$ (which makes sense, for a.e. line ℓ, and is measurable, by Fubini) is absolutely continuous, which means that it is differentiable almost everywhere on $\ell \cap \Omega$ and is the indefinite integral of its derivative on each component of $\ell \cap \Omega$. By the natural identification, the derivative in question is obtained from the distributional gradient of u.

Lemma 3.1. *Every $u \in W$ is absolutely continuous on lines in Ω.*

Proof. This lemma can be seen as a consequence of [**Maz**, Theorem 1.1.3/1] since the absolute continuity on lines is a local property and, thanks to (2.13), $W \subset \{u \in L^1_{loc}(\Omega),\, \nabla u \in L^2_{loc}(\Omega)\}$. Yet, the proof of Lemma 3.1 is classical: since the property is local, it is enough to check the property on lines parallel to a fixed vector e, and when Ω is the product of n intervals, one of which is parallel to e. This last amounts to using the definition of the distributional gradient, testing on product functions, and applying Fubini. In addition, the derivative of u on almost every line ℓ of direction e coincides with $\nabla u \cdot e$ almost everywhere on ℓ. □

Lemma 3.2. *We have the following equality of spaces*

$$W = \{u \in L^1_{loc}(\mathbb{R}^n),\, \nabla u \in L^2(\mathbb{R}^n, w)\}, \tag{3.2}$$

where the derivative of u is taken in the sense of distribution in $\mathbb{R}^n$, that is for any $\varphi \in C_0^\infty(\mathbb{R}^n, \mathbb{R}^n)$,

$$\int \nabla u \cdot \varphi \, dx = -\int u \operatorname{div} \varphi \, dx.$$

PROOF. Here and in the sequel, we will constantly use the fact that with $\Omega = \mathbb{R}^n \setminus \Gamma$ and because (1.1) holds with $d < n-1$,

$$\text{almost every line } \ell \text{ is contained in } \Omega. \tag{3.3}$$

Let us recall that it means that given any choice of direction v and a vector hyperplane P transverse to v, the line $x + \mathbb{R}v \subset \Omega$ for almost every $x \in P$. In particular, for almost every $(x, y) \in (\mathbb{R}^n)^2$, there is a unique line going through x and y and this line is included in Γ.

Lemma 3.1 and (3.3) implies that $u \in W$ is actually absolutely continuous on lines in $\mathbb{R}^n$, i.e. any $u \in W$ (possibly modified on a set of zero measure) is absolutely continuous on almost every line $\ell \subset \mathbb{R}^n$. As we said before, $\nabla u = (\partial_1 u, \dots, \partial_n u)$, the distributional gradient of u in Ω, equals the 'classical' gradient of u defined in the following way. If $e_1 = (1, 0, \dots, 0)$ is the first coordinate vector, then $\partial_1 u(y, z)$ is the derivative at the point y of the function $u_{|(0,z)+\mathbb{R}e_1}$, the latter quantity being defined for almost every $(y, z) \in \mathbb{R} \times \mathbb{R}^{n-1}$ because u is absolutely continuous on lines in $\mathbb{R}^n$. If $i > 1$, $\partial_i u(x)$ is defined in a similar way.

As a consequence, for almost any $(y, z) \in \mathbb{R}^n \times \mathbb{R}^n$, $u(z) - u(y) = \int_0^1 (z - y) \cdot \nabla u(y + t(z - y))dt$ and hence,

$$|u(y) - u(z)| \leq \int_0^1 |z - y||\nabla u(y + t(z - y))|dt. \tag{3.4}$$

Let us integrate this for y in a ball B. We get that for almost every $z \in \mathbb{R}^n$,

$$\fint_{y \in B} |u(y) - u(z)|dy \leq \fint_{y \in B} \int_0^1 |z - y||\nabla u(y + t(z - y))|dt. \tag{3.5}$$

Let us further restrict to the case $z \in B = B(x, r)$; the change of variable $\xi = z + t(y - z)$ shows that

$$\begin{aligned} \fint_{y \in B} |u(y) - u(z)|dy &= \int_0^1 \fint_{y \in B} |y - z||\nabla u(z + t(y - z))|dy\, dt \\ &= \int_0^1 \frac{1}{|B|} \int_{\xi \in B(z+t(x-z),tr)} \frac{|z - \xi|}{t} |\nabla u(\xi)| \frac{d\xi}{t^n} dt \\ &= \int_{\xi \in B} |\nabla u(\xi)| \frac{|z - \xi|}{|B(z, r)|} d\xi \int_{|z-\xi|/2r}^1 \frac{dt}{t^{n+1}} \\ &\leq 2^n |B(0, 1)|^{-1} \int_{\xi \in B} |\nabla u(\xi)||z - \xi|^{1-n} d\xi, \end{aligned} \tag{3.6}$$

where the last but one line is due to the fact that $\xi \in B(z+t(x-z), tr)$ is equivalent to $|\xi - z - t(x - z)| \leq tr$, which forces $|\xi - z| \leq tr + t|x - z| \leq 2rt$. Therefore, for almost any $z \in B$,

$$\fint_{y \in B} |u(y) - u(z)|dy \leq C \int_{\xi \in B} |\nabla u(\xi)||z - \xi|^{1-n} d\xi, \tag{3.7}$$

where C depends on n, but not on r, u, or z. With a second integration on $z \in B = B(x,r)$, we obtain

$$\begin{aligned}(3.8)\qquad \fint_{z\in B}\fint_{y\in B}|u(y)-u(z)|dy\,dz &\leq C\int_{\xi\in B}|\nabla u(\xi)|\fint_{z\in B}|z-\xi|^{1-n}dz\,d\xi \\ &\leq Cr\fint_{\xi\in B}|\nabla u(\xi)|d\xi.\end{aligned}$$

By Hölder's inequality and (2.13), the right-hand side is bounded (up to a constant depending on r) by $\|u\|_W$. As a consequence,

$$(3.9)\qquad \fint_{z\in B}\fint_{y\in B}|u(y)-u(z)| \leq C_r\|u\|_W < +\infty.$$

and thus, by Fubini's lemma, $\fint_{y\in B}|u(y)-u(z)| < +\infty$ for a.e. $z \in B$. In particular, the quantity $\fint_{y\in B}|u(y)|$ is finite for any ball $B \subset \mathbb{R}^n$, that is $u \in L^1_{loc}(\mathbb{R}^n)$.

Since $L^1_{loc}(\mathbb{R}^n) \subset L^1_{loc}(\Omega)$, we just proved that $W = \{u \in L^1_{loc}(\mathbb{R}^n),\ \nabla u \in L^2(\Omega,w)\}$, where $\nabla u = (\partial_1 u,\dots,\partial_n u)$ is distributional gradient on Ω. Let $u \in W$. Since Γ has zero measure, $\nabla u \in L^2(\mathbb{R}^n,w)$ and thus it suffices to check that u has actually a distributional derivative in $\mathbb{R}^n$ and that this derivative equals ∇u. However, the latter fact is a simple consequence of [**Maz**, Theorem 1.1.3/2], because u is absolutely continuous on lines in $\mathbb{R}^n$. The proof of Maz'ya's result is basically the following: for any $i \in \{1,\dots,n\}$ and any $\phi \in C_0^\infty(\mathbb{R}^n)$, an integration by part gives $\int u\partial_i\phi = -\int(\partial_i u)\phi$. The two integrals in the latter equality make sense since both u and $\partial_i u$ are in $L^1_{loc}(\mathbb{R}^n)$; the integration by part is possible because u is absolutely continuous on almost every line. □

REMARK 3.3. An important by-product of the proof is that Lemma 3.1 can be improved into: for any $u \in W$ (possibly modified on a set of zero measure) and almost every line $\ell \subset \mathbb{R}^n$, $u_{|\ell}$ is absolutely continuous. This property will be referred to as (ACL).

THEOREM 3.4. *There exists a bounded linear operator $T : W \to H$ (a trace operator) with the following properties. The trace of $u \in W$ is such that, for σ-almost every $x \in \Gamma$,*

$$(3.10)\qquad Tu(x) = \lim_{r\to 0}\fint_{B(x,r)} u(y)dy := \lim_{r\to 0}\frac{1}{|B(x,r)|}\int u(y)dy$$

and, analogously to the Lebesgue density property,

$$(3.11)\qquad \lim_{r\to 0}\fint_{B(x,r)}|u(y)-Tu(x)|dy = 0.$$

PROOF. First, we want bounds on ∇u near $x \in \Gamma$, so we set

$$(3.12)\qquad M_r(x) = \fint_{B(x,r)}|\nabla u|^2\,dz$$

and estimate $\int_\Gamma M_r(x)d\sigma(x)$. We cover Γ by balls $B_j = B(x_j,r)$ centered on Γ such that the $2B_j = B(x_j,2r)$ have bounded overlap (we could even make the $B(x_j,r/5)$ disjoint), and notice that for $x \in B_j$,

$$(3.13)\qquad M_r(x) \leq Cr^{-n}\int_{2B_j}|\nabla u|^2\,dz.$$

We sum and get that

$$\int_\Gamma M_r(x)d\sigma(x) \le \sum_j \int_{B_j} M_r(x)d\sigma(x) \le C\sum_j \sigma(B_j) \sup_{x\in B_j} M_r(x)$$
$$\le C\sum_j \sigma(B_j) r^{-n} \int_{2B_j} |\nabla u|^2\, dz \le Cr^{d-n}\sum_j \int_{2B_j} |\nabla u|^2\, dz$$
$$\le Cr^{d-n}\int_{\Gamma(2r)} |\nabla u|^2\, dz \tag{3.14}$$

because the $2B_j$ have bounded overlap and where $\Gamma(2r)$ denotes a $2r$-neighborhood of Γ. Next set

$$N(x) = \sum_{k\ge 0} 2^{-k} M_{2^{-k}}(x); \tag{3.15}$$

then

$$\int_\Gamma N(x)d\sigma(x) = \sum_{k\ge 0} 2^{-k}\int_\Gamma M_{2^{-k}}(x)d\sigma(x) \le C\sum_{k\ge 0} 2^{k(n-d-1)}\int_{\Gamma(2^{-k+1})} |\nabla u(z)|^2 dz$$
$$\le C\int_{\Gamma(2)} |\nabla u(z)|^2 a(z)dz, \tag{3.16}$$

where $a(z) = \sum_{k\ge 0} 2^{k(n-d-1)}\mathbb{1}_{z\in\Gamma(2^{-k+1})}$. For a given $z\in\Omega$, $z\in\Gamma(2^{-k+1})$ only for k so small that $\delta(z)\le 2^{-k+1}$. The largest values of $2^{k(n-d-1)}$ are for k as large as possible, when $2^{-k}\approx\delta(z)$; thus $a(z)\le C\delta(z)^{-n+d+1} = Cw(z)$, and

$$\int_\Gamma N(x)d\sigma(x) \le C\int_{\Gamma(2)} |\nabla u(z)|^2 w(z)dz. \tag{3.17}$$

Our trace function $g = Tu$ will be defined as the limit of the functions g_r, where

$$g_r(x) = \fint_{z\in B(x,r)} u(z)dz. \tag{3.18}$$

Our aim is to use the estimates established in the proof of Lemma 3.2. Notice that for $x\in\Gamma$ and $r>0$,

$$\fint_{z\in B(x,r)} |u(z)-g_r(x)|dz = \fint_{z\in B(x,r)} \left|u(z) - \fint_{\xi\in B(x,r)} u(y)dy\right| dz$$
$$\le \fint_{z\in B(x,r)}\fint_{y\in B(x,r)} \Big|u(z)-u(y)\Big| dy\, dz. \tag{3.19}$$

By (3.8),

$$\fint_{z\in B(x,r)} |u(z)-g_r(x)|dz \le \fint_{z\in B(x,r)}\fint_{y\in B(x,r)} \Big|u(z)-u(y)\Big| dy\, dz$$
$$\le Cr^{-n+1}\int_{\xi\in B(x,r)} |\nabla u(\xi)|d\xi. \tag{3.20}$$

Thus for $r/10\le s\le r$,

$$|g_s(x)-g_r(x)| = \left|\fint_{z\in B(x,s)} u(z)dz - g_r(x)\right| \le \fint_{z\in B(x,s)} |u(z)-g_r(x)|dz$$
$$\le Cr\fint_{\xi\in B(x,r)} |\nabla u(\xi)|d\xi \le CrM_r(x)^{1/2}. \tag{3.21}$$

Set $\Delta_r(x) = \sup_{r/10\le s\le r} |g_s(x) - g_r(x)|$; we just proved that $\Delta_r(x) \le CrM_r(x)^{1/2}$. Let $\alpha \in (0, 1/2)$ be given. If $N(x) < +\infty$, we get that

$$\sum_{k\ge 0} 2^{\alpha k}\Delta_{2^{-k}}(x) \le C\sum_{k\ge 0} 2^{\alpha k}2^{-k}M_{2^{-k}}(x)^{1/2}$$

$$\le C\Big\{\sum_{k\ge 0} 2^{-k}M_{2^{-k}}(x)\Big\}^{1/2}\Big\{\sum_{k\ge 0} 2^{2\alpha k}2^{-k}\Big\}^{1/2} \le CN(x)^{1/2} < +\infty. \tag{3.22}$$

Therefore, $\sum_{k\ge 0}\Delta_{2^{-k-2}}(x)$ converges (rather fast), and since (3.17) implies that $N(x) < +\infty$ for σ-almost every $x \in \Gamma$, it follows that there exists

$$g(x) = \lim_{r\to 0} g_r(x) \quad \text{for } \sigma\text{-almost every } x \in \Gamma. \tag{3.23}$$

In addition, we may integrate (the proof of) (3.22) and get that for $2^{-j-1} < r \le 2^{-j}$,

$$\begin{aligned} \|g - g_r\|^2_{L^2(\sigma)} &= \int_\Gamma |g(x) - g_r(x)|^2 d\sigma(x) \le \int_\Gamma \Big\{\sum_{k\ge j}\Delta_{2^{-k}}(x)\Big\}^2 d\sigma(x) \\ &\le C2^{-2\alpha j}\int_\Gamma \Big\{\sum_{k\ge j} 2^{\alpha j}\Delta_{2^{-k}}(x)\Big\}^2 d\sigma(x) \\ &\le Cr^{2\alpha}\int_\Gamma N(x)d\sigma(x) \le Cr^{2\alpha}\|u\|^2_W \end{aligned} \tag{3.24}$$

by (3.23) and the definition of $\Delta_r(x)$, then (3.22) and (3.17). Thus g_r converges also (rather fast) to g in L^2. Let us make an additional remark. Fix $r > 0$ and $\alpha \in (0, 1/2)$. For any ball B centered on Γ,

$$\|g\|_{L^1(B,\sigma)} \le C_B\|g - g_r\|_{L^2(\sigma)} + \|g_r\|_{L^1(B,\sigma)} \tag{3.25}$$

by Hölder's inequality. The first term is bounded with (3.24). Use (1.1) and Fubini's theorem to bound the second one by $C_r\|g\|_{L^1(\widetilde{B})}$, where $\widetilde{B}$ is a large ball (that depends on r and contains B). As a consequence,

$$\text{for any } u \in W,\ g = Tu \in L^1_{loc}(\Gamma, \sigma). \tag{3.26}$$

This completes the definition of the trace $g = T(u)$. We announced (as a Lebesgue property) that

$$\lim_{r\to 0} \fint_{B(x,r)} |u(y) - Tu(x)|dy \tag{3.27}$$

for σ-almost every $x \in \Gamma$, and indeed

$$\begin{aligned} &\fint_{B(x,r)} |u(y) - Tu(x)|dy \\ &\quad = \fint_{B(x,r)} |u(y) - g(x)|dy \le |g(x) - g_r(x)| + \fint_{B(x,r)} |u(y) - g_r(x)|\,dy \\ &\quad \le |g(x) - g_r(x)| + Cr\fint_{B(x,3r)} |\nabla u|\,dy \le |g(x) - g_r(x)| + CrM_{4r}(x)^{1/2} \end{aligned} \tag{3.28}$$

by (3.20) and the second part of (3.21). The first part tends to 0 for σ-almost every $x \in \Gamma$, by (3.23), and the second part tends to 0 as well, because $N(x) < +\infty$ almost everywhere and by the definition (3.15).

Next we show that $g = Tu$ lies in the Sobolev space $H = H^{1/2}(\Gamma)$, i.e., that

$$\|g\|_H^2 = \int_\Gamma \int_\Gamma \frac{|g(x) - g(y)|^2}{|x-y|^{d+1}} d\sigma(x) d\sigma(y) < +\infty. \tag{3.29}$$

The simplest will be to prove uniform estimates on the g_r, and then go to the limit. Let us fix $r > 0$ and consider the integral

$$I(r) = \int_{x\in\Gamma} \int_{y\in\Gamma; |y-x|\geq r} \frac{|g_r(x) - g_r(y)|^2}{|x-y|^{d+1}} d\sigma(x) d\sigma(y). \tag{3.30}$$

Set $Z_k(r) = \{(x,y) \in \Gamma \times \Gamma\,;\, 2^k r \leq |y-x| < 2^{k+1} r\}$ and

$$I_k(r) = \int\int_{Z_k(r)} \frac{|g_r(x) - g_r(y)|^2}{|x-y|^{d+1}} d\sigma(x) d\sigma(y).$$

Thus $I(r) = \sum_{k\geq 0} I_k(r)$ and

$$I_k(r) \leq (2^k r)^{-d-1} \int\int_{Z_k(r)} |g_r(x) - g_r(y)|^2 d\sigma(x) d\sigma(y). \tag{3.31}$$

Fix $k \geq 0$, set $\rho = 2^{k+1} r$, and observe that for $(x,y) \in Z_k(r)$,

$$\begin{aligned}
&|g_r(x) - g_\rho(y)| \\
&\quad = \Big| \fint_{z\in B(x,r)} \fint_{\xi\in B(y,\rho)} [u(z) - u(\xi)] dz d\xi \Big| \\
&\quad \leq \fint_{z\in B(x,r)} \fint_{\xi\in B(y,\rho)} |u(z) - u(\xi)| d\xi dz \\
&\quad \leq 3^n \fint_{z\in B(x,r)} \Big\{ \fint_{\xi\in B(z,3\rho)} |u(z) - u(\xi)| d\xi \Big\} dz \\
&\quad \leq C\rho^n \fint_{z\in B(x,r)} \fint_{\zeta\in B(z,3\rho)} |\nabla u(\zeta)| |z-\zeta|^{1-n} d\zeta dz
\end{aligned} \tag{3.32}$$

because $B(y,\rho) \subset B(z,3\rho)$ and by (3.7). We apply Cauchy-Schwarz, with an extra bit $|z-\zeta|^{-\alpha}$, where $\alpha > 0$ will be taken small, and which will be useful for convergence later

$$\begin{aligned}
&|g_r(x) - g_\rho(y)|^2 \\
&\leq C\rho^{2n} \Big\{ \fint_{z\in B(x,r)} \fint_{\zeta\in B(z,3\rho)} |\nabla u(\zeta)|^2 |z-\zeta|^{1-n+\alpha} \Big\} \\
&\qquad\qquad \Big\{ \fint_{z\in B(x,r)} \fint_{\zeta\in B(z,3\rho)} |z-\zeta|^{1-n-\alpha} \Big\} \\
&\leq C\rho^{n+1-\alpha} \fint_{z\in B(x,r)} \fint_{\zeta\in B(z,3\rho)} |\nabla u(\zeta)|^2 |z-\zeta|^{1-n+\alpha} d\zeta dz.
\end{aligned} \tag{3.33}$$

The same computation, with $g_r(y)$, yields

$$|g_r(y) - g_\rho(y)|^2 \leq C\rho^{n+1-\alpha} \fint_{z\in B(y,r)} \fint_{\zeta\in B(z,3\rho)} |\nabla u(\zeta)|^2 |z-\zeta|^{1-n+\alpha} d\zeta dz. \tag{3.34}$$

We add the two and get an estimate for $|g_r(x)-g_r(y)|^2$, which we can integrate to get that

$$
\begin{aligned}
& I_k(r) \\
& \leq C\rho^{-d-1}\rho^{n+1-\alpha}\int\int_{(x,y)\in Z_k(r)}\fint_{z\in B(x,r)}\fint_{\zeta\in B(z,3\rho)}\frac{|\nabla u(\zeta)|^2}{|z-\zeta|^{1-n+\alpha}}d\zeta dz d\sigma(x)d\sigma(y) \\
& \leq C\rho^{-d-\alpha}r^{-n}\int\int_{(x,y)\in Z_k(r)}\int_{z\in B(x,r)}\int_{\zeta\in B(z,3\rho)}\frac{|\nabla u(\zeta)|^2}{|z-\zeta|^{1-n+\alpha}}d\zeta dz d\sigma(x)d\sigma(y)
\end{aligned} \tag{3.35}
$$

by (3.31), (3.33), and (3.34), and where we can drop the part that comes from (3.34) by symmetry. We integrate in $y\in\Gamma$ such that $2^k r\leq |x-y|\leq 2^{k+1}r$ and get that

$$
\begin{aligned}
I_k(r) &\leq C\rho^{-\alpha}r^{-n}\int_{x\in\Gamma}\int_{z\in B(x,r)}\int_{\zeta\in B(z,3\rho)}|\nabla u(\zeta)|^2|z-\zeta|^{1-n+\alpha}d\zeta dz d\sigma(x) \\
&\leq C\int_{\zeta\in\Omega}|\nabla u(\zeta)|^2 h_k(\zeta)d\zeta,
\end{aligned} \tag{3.36}
$$

with

$$
h_k(\zeta)=\rho^{-\alpha}r^{-n}\int_{x\in\Gamma}\int_{z\in B(x,r)\cap B(\zeta,3\rho)}|z-\zeta|^{1-n+\alpha}dz d\sigma(x). \tag{3.37}
$$

We start with the contribution $h_k^0(\zeta)$ of the region where $|x-\zeta|\geq 2r$, where the computation is simpler because $|z-\zeta|\geq\frac{1}{2}|x-\zeta|$ there. We get that

$$
\begin{aligned}
h_k^0(\zeta) &\leq C\rho^{-\alpha}r^{-n}\int_{x\in\Gamma}\int_{z\in B(x,r)\cap B(\zeta,3\rho)}|x-\zeta|^{1-n+\alpha}dz d\sigma(x) \\
&\leq C\rho^{-\alpha}\int_{x\in\Gamma\cap B(\zeta,4\rho)}|x-\zeta|^{1-n+\alpha}d\sigma(x).
\end{aligned} \tag{3.38}
$$

With ζ, r, and ρ fixed, $h_k^0(\zeta)$ vanishes unless $\delta(\zeta)=\mathrm{dist}(\zeta,\Gamma)<4\rho$. The region where $|x-\zeta|$ is of the order of $2^m\delta(\zeta)$, $m\geq 0$, contributes less than $C(2^m\delta(\zeta))^{d+1-n+\alpha}$ to the integral (because σ is Ahlfors-regular). If α is chosen small enough, the exponent is still negative, the largest contribution comes from $m=0$, and $h_k^0(\zeta)\leq C\rho^{-\alpha}\delta(\zeta)^{d+1-n+\alpha}$. Recall that $\rho=2^{k+1}r$, and k is such that $\delta(\zeta)<4\rho$; we sum over k and get that

$$
\sum_k h_k^0(\zeta)\leq C\sum_{k\geq 0\,;\,\delta(\zeta)<4\rho}\rho^{-\alpha}\delta(\zeta)^{d+1-n+\alpha}\leq C\delta(\zeta)^{d+1-n}, \tag{3.39}
$$

because this time the smallest values of ρ give the largest contributions. We are left with

(3.40)

$$
h_k^1(\zeta)=h_k(\zeta)-h_k^0(\zeta)=\rho^{-\alpha}r^{-n}\int_{x\in\Gamma\cap B(\zeta,2r)}\int_{z\in B(x,r)\cap B(\zeta,3\rho)}|z-\zeta|^{1-n+\alpha}dz d\sigma(x).
$$

Notice that $|z-\zeta|\leq|z-x|+|x-\zeta|\leq 3r$; we use the local Ahlfors-regularity to get rid of the integral on Γ, and get that

$$
h_k^1(\zeta)\leq C\rho^{-\alpha}r^{-n}r^d\int_{z\in B(\zeta,3r)}|z-\zeta|^{1-n+\alpha}dz\leq C\rho^{-\alpha}r^{d+1-n+\alpha}. \tag{3.41}
$$

We sum over k and get that $\sum_k h_k^1(\zeta) \leq Cr^{d+1-n} \leq C\delta(\zeta)^{d+1-n}$, because if $\delta(\zeta) \geq 2r$, we simply get that $h_k^1(\zeta) = 0$ for all k, because $\Gamma \cap B(\zeta, 2r) = \emptyset$ and by (3.40). Altogether, $\sum_k h_k(\zeta) \leq C\delta(\zeta)^{d+1-n}$, and

$$I(r) = \sum_k I_k(r) \leq C \sum_k \int_{\zeta \in \Omega} |\nabla u(\zeta)|^2 h_k(\zeta) d\zeta \\ \leq C \int_{\zeta \in \Omega} |\nabla u(\zeta)|^2 \delta(\zeta)^{d+1-n} d\zeta = C\|u\|_W^2 \tag{3.42}$$

by definition of the $I_k(r)$, then (3.36) and the definition of W. We may now look at the definition (3.30) of $I(r)$, let r tend to 0, and get that

$$\|g\|_H \leq C\|u\|_W^2 \tag{3.43}$$

by Fatou's lemma, as needed for the trace theorem. □

CHAPTER 4

Poincaré Inequalities

LEMMA 4.1. *Let Γ be a d-ADR set in $\mathbb{R}^n$, $d < n-1$, that is, assume that* (1.1) *is satisfied. Then*

$$\fint_{B(x,r)} |u(y)|dy \leq Cr^{-d} \int_{B(x,r)} |\nabla u(y)|w(y)dy \tag{4.1}$$

for $u \in W$, $x \in \Gamma$, and $r > 0$ such that $Tu = 0$ on $\Gamma \cap B(x,r)$.

PROOF. To simplify the notation we assume that $x = 0$.

We should of course observe that the right-hand side of (4.1) is finite. Indeed, recall that Lemma 2.3 gives

$$\int_{\xi \in B(0,r)} w(\xi)d\xi \leq Cr^{1+d}; \tag{4.2}$$

then by Cauchy-Schwarz

$$\begin{aligned} r^{-d} &\int_{\xi \in B(0,r)} |\nabla u(\xi)|w(\xi)d\xi \\ &\leq r^{-d}\Big\{ \int_{\xi \in B(0,r)} |\nabla u(\xi)|^2 w(\xi)\, d\xi \Big\}^{1/2} \Big\{ \int_{\xi \in B(0,r)} w(\xi)\, d\xi \Big\}^{1/2} \\ &\leq r^{\frac{1-d}{2}} \Big\{ \int_{\xi \in B(0,r)} |\nabla u(\xi)|^2 w(\xi)\, d\xi \Big\}^{1/2}. \end{aligned} \tag{4.3}$$

The homogeneity still looks a little weird because of the weight (but things become simpler if we think that $\delta(\xi)$ is of the order of r), but at least the right-hand side is finite because $u \in W$.

Turning to the proof of (4.1), to avoid complications with the fact that (3.4) and (3.5) do not necessarily hold σ-almost everywhere on Γ, let us use the g_s again. We first prove that for $s < r$ small,

$$\fint_{y \in B(0,r)} \fint_{x \in \Gamma \cap B(0,r/2)} |u(y) - g_s(x)|d\sigma(x)\, dy \leq Cr^{-d} \fint_{B(0,r)} |\nabla u(\xi)|\delta(\xi)^{1+d-n} d\xi. \tag{4.4}$$

Denote by $I(s)$ the left-hand side. By (3.18),

$$I(s) \leq \fint_{y \in B(0,r)} \fint_{x \in \Gamma \cap B(0,r/2)} \fint_{z \in B(x,s)} |u(y) - u(z)|dz\, d\sigma(x)\, dy. \tag{4.5}$$

For x fixed, we can still prove as in (3.7) that

$$\fint_{y \in B(0,r)} |u(y) - u(z)|dy \leq C \int_{B(0,r)} |\nabla u(\xi)||z - \xi|^{1-n} d\xi \tag{4.6}$$

(for $x \in \Gamma \cap B(0,r/2)$ and $z \in B(x,s)$, there is even a bilipschitz change of variable that sends z to 0 and maps $B(0,r)$ to itself). We are left with

$$I(s) \le C \fint_{x\in\Gamma\cap B(0,r/2)} \fint_{z\in B(x,s)} \int_{\xi\in B(0,r)} |\nabla u(\xi)||z-\xi|^{1-n} d\xi dz d\sigma(x). \tag{4.7}$$

The main piece of the integral will again be called $I_0(s)$, where we integrate in the region where $|\xi - x| \ge 2s$ and hence $|z-\xi|^{1-n} \le 2^n|x-\xi|^{1-n}$. Thus

$$\begin{aligned} I_0(s) &\le C \int_{\xi\in B(0,r)} \fint_{x\in\Gamma\cap B(0,r/2)} \fint_{z\in B(x,s)} |\nabla u(\xi)||x-\xi|^{1-n} dz d\sigma(x) d\xi \\ &\le Cr^{-d} \int_{\xi\in B(0,r)} \int_{x\in\Gamma\cap B(0,r/2)} |\nabla u(\xi)||x-\xi|^{1-n} d\sigma(x) d\xi \\ &\le Cr^{-d} \int_{\xi\in B(0,r)\setminus\Gamma} |\nabla u(\xi)| h(\xi) d\xi, \end{aligned} \tag{4.8}$$

where for $\xi \in B(0,r)\setminus\Gamma$ we set

$$h(\xi) = \int_{x\in\Gamma\cap B(0,r/2)} |x-\xi|^{1-n} d\sigma(x) \le C\delta(\xi)^{1-n+d} \tag{4.9}$$

where for the last inequality we cut the domain of integration into pieces where $|x-\xi| \approx 2^m\delta(\xi)$ and use (1.1). For the other piece of (4.7) where $|\xi - x| < 2s$, we get the integral

$$\begin{aligned} &I_1(s) \\ &\le Cr^{-d}s^{-n} \int_{\xi\in B(0,r)} \int_{x\in\Gamma\cap B(0,r/2)\cap B(\xi,2s)} \int_{z\in B(x,s)} |\nabla u(\xi)||z-\xi|^{1-n} d\xi dz d\sigma(x) \\ &\le Cr^{-d}s^{d-n} \int_{\xi\in B(0,r);\delta(\xi)\le 2s} \int_{z\in B(\xi,3s)} |\nabla u(\xi)||z-\xi|^{1-n} d\xi dz \\ &\le Cr^{-d}s^{1+d-n} \int_{\xi\in B(0,r);\delta(\xi)\le 2s} |\nabla u(\xi)| d\xi \le Cr^{-d} \int_{\xi\in B(0,r)} |\nabla u(\xi)|\delta(\xi)^{1+d-n} d\xi. \end{aligned} \tag{4.10}$$

Altogether

$$I(s) \le Cr^{-d} \int_{\xi\in B(0,r)} |\nabla u(\xi)|\delta(\xi)^{1+d-n} d\xi, \tag{4.11}$$

which is (4.4). When s tends to 0, $g_s(x)$ tends to $g(x) = Tu(x) = 0$ for σ-almost every $x \in \Gamma \cap B(0,r/2)$, and we get (4.1) by Fatou. □

LEMMA 4.2. *Let Γ be a d-ADR set in $\mathbb{R}^n$, $d < n-1$, that is, assume that (1.1) is satisfied. Let $p \in \left[1, \frac{2n}{n-2}\right]$ (or $p \in [1,+\infty)$ if $n = 2$). Then for any $u \in W$, $x \in \mathbb{R}^n$ and $r > 0$*

$$\begin{aligned} &\Big\{\frac{1}{m(B(x,r))} \int_{B(x,r)} \left|u(y) - u_{B(x,r)}\right|^p w(y) dy\Big\}^{1/p} \\ &\le Cr\Big\{\frac{1}{m(B(x,r))} \int_{B(x,r)} |\nabla u(y)|^2 w(y) dy\Big\}^{1/2}, \end{aligned} \tag{4.12}$$

where u_B denotes either $\fint_B u\, dz$ or $m(B)^{-1}\int_B u\, w(z) dz$. If $x \in \Gamma$ and, in addition, $Tu = 0$ on $\Gamma \cap B(x,r)$ then

$$\Big\{r^{-d-1} \int_{B(x,r)} |u(y)|^p w(y) dy\Big\}^{1/p} \le Cr\Big\{r^{-d-1} \int_{B(x,r)} |\nabla u(y)|^2 w(y) dy\Big\}^{1/2}. \tag{4.13}$$

PROOF. In the proof, we will use $dm(z)$ for $w(z)dz$ and hence, for instance $\int_B u\,dm$ denotes $\int_B u(z)w(z)dz$. We start with the following inequality. Let $p \in [1,+\infty)$. If $u \in L^p_{loc}(\mathbb{R}^n, w) \subset L^1_{loc}(\mathbb{R}^n)$, then for any ball B,

$$\int_B \left| u - \fint_B u\,dy \right|^p dm \approx \int_B \left| u - \frac{1}{m(B)} \int_B u\,dm \right|^p dm. \tag{4.14}$$

First we bound the left-hand side. We introduce $m(B)^{-1}\int_B u\,dm$ inside the absolute values and then use the triangle inequality:

$$\begin{aligned}
&\int_B \left| u - \fint_B u\,dy \right| dm \\
&\quad \le C\int_B \left| u(z) - \frac{1}{m(B)} \int_B u\,dm \right|^p dm + Cm(B) \left| \fint_B u - \frac{1}{m(B)} \int_B u\,dm \right|^p \\
&\quad \le C\int_B \left| u(z) - \frac{1}{m(B)} \int_B u\,dm \right|^p dm + C\frac{m(B)}{|B|} \int_B \left| u - \frac{1}{m(B)} \int_B u\,dm \right|^p \\
&\quad \le C\int_B \left| u(z) - \frac{1}{m(B)} \int_B u\,dm \right|^p dm,
\end{aligned} \tag{4.15}$$

where the last line is due to (2.13). The reverse estimate is quite immediate

$$\begin{aligned}
&\int_B \left| u - \frac{1}{m(B)} \int_B u\,dm \right|^p dm \\
&\quad \le C\int_B \left| u(z) - \fint_B u\,dy \right|^p dm + Cm(B) \left| \frac{1}{m(B)} \int_B u\,dm - \fint_B u\,dy \right|^p \\
&\quad \le C\int_B \left| u(z) - \fint_B u\,dy \right|^p dm + Cm(B) \left| \frac{1}{m(B)} \int_B \left(u - \fint_B u\,dy \right) dm \right|^p \\
&\quad \le C\int_B \left| u(z) - \fint_B u\,dy \right|^p dm,
\end{aligned} \tag{4.16}$$

which finishes the proof of (4.14).

In the sequel of the proof, we write u_B for $m(B)^{-1}\int_B u\,dm$. Thanks to (4.14), it suffices to prove (4.12) only for this particular choice of u_B. We now want to prove a (1,1) Poincaré inequality, that is

$$\int_B |u(z) - u_B|\,dm \le Cr \int_B |\nabla u(z)| dm. \tag{4.17}$$

for any $u \in W$ and any ball $B \subset \mathbb{R}^n$ of radius r. In particular, $u \in L^1_{loc}(\mathbb{R}^n, w)$.

Let $B \subset \mathbb{R}^n$ of radius r. Recall first that thanks to Lemma 3.2, $\fint_B u$ makes sense for every ball B. If we prove for $u \in W$ the estimate

$$\int_B \left| u - \fint_B u\,dy \right| dm \le Cr \int_B |\nabla u| dm, \tag{4.18}$$

then (4.17) will follows. Indeed, assume (4.18) holds for any ball B. The left-hand of (4.18) is then bounded, up to a constant, by $r\|u\|_W$ and is thus finite. Therefore, for any ball B, $\int_B |u|dm \le \int_B |u - \fint_B u|dm + |\fint_B u\,dy| < +\infty$, i.e. $u \in L^1_{loc}(\mathbb{R}^n, w)$. But now, $u \in L^1_{loc}(\mathbb{R}^n, w)$, so we can use (4.14). Together with (4.18), it implies (4.17).

We want to prove (4.18). The estimate (3.7) yields

$$\begin{aligned}\int_B \left| u - \fint_B u\,dy \right| dm &\leq C \int_B \int_B |\nabla u(\xi)||z-\xi|^{1-n} w(z) d\xi\, dz \\ &\leq C \int_B |\nabla u(\xi)| \int_{B(\xi,2r)} |z-\xi|^{1-n} w(z) dz\, d\xi\end{aligned} \tag{4.19}$$

and thus it remains to check that for $\xi \in \mathbb{R}^n$ and $r > 0$,

$$I = \int_{B(\xi,2r)} |z-\xi|^{1-n} w(z) dz \leq C r w(\xi). \tag{4.20}$$

First, note that if $\delta(\xi) \geq 4r$, then for all $z \in B(\xi, 2r)$, one has

$$\frac{1}{2}\delta(\xi) \leq \delta(\xi) - 2r \leq \delta(z) \leq \delta(\xi) + 2r \leq 2\delta(\xi)$$

and so $w(z)$ is equivalent to $w(\xi)$. Thus $I \leq Cw(\xi) \int_{B(\xi,2r)} |z-\xi|^{1-n} dz \leq Crw(\xi)$. It remains to prove the case $\delta(\xi) < 4r$. We split I into $I_1 + I_2$ where, for I_1, the domain of integration is restrained to $B(\xi, \delta(\xi)/2)$. For any $z \in B(\xi, \delta(\xi)/2)$, we have $w(z) \leq Cw(\xi)$ and thus

$$I_1 \leq Cw(\xi) \int_{B(\xi,\delta(\xi)/2)} |z-\xi|^{1-n} dz \leq Cw(\xi)\delta(\xi) \leq Crw(\xi). \tag{4.21}$$

It remains to bound I_2. In order to do it, we decompose the remaining domain into annuli $C_j(\xi) := \{z \in \mathbb{R}^n,\, 2^{j-1}\delta(\xi) \leq |\xi - z| \leq 2^j \delta(\xi)\}$. We write κ for the smallest integer bigger than $\log_2(2r/\delta(\xi))$, which is the highest value for which $C_\kappa \cap B(\xi, 2r)$ is non-empty. We have

(4.22)

$$I_2 \leq C \sum_{j=0}^{\kappa} 2^{j(1-n)} \delta(\xi)^{1-n} \int_{C_j(\xi)} dm(z) \leq C \sum_{j=0}^{\kappa} 2^{j(1-n)} \delta(\xi)^{1-n} m(B(\xi, 2^j \delta(\xi))).$$

The ball $B(\xi, 2^j\delta(\xi))$ is close to Γ and thus Lemma 2.3 gives that the quantity $m(B(\xi, 2^j\delta(\xi)))$ is bounded, up to a constant, by $2^{j(d+1)}\delta(x)^{d+1}$. We deduce, since $2 + d - n \leq 1$, that

$$\begin{aligned} I_2 &\leq C \sum_{j=0}^{\kappa} 2^{j(2+d-n)} \delta(\xi)^{2+d-n} \leq C\delta(\xi)^{2+d-n} \sum_{j=0}^{\kappa} 2^j \\ &\leq C\delta(\xi)^{2+d-n} \left(\frac{2r}{\delta(\xi)}\right) \leq Cr\delta(\xi)^{1+d-n} = Crw(\xi), \end{aligned} \tag{4.23}$$

which ends the proof of (4.20) and thus also the one of the Poincaré inequality (4.17).

Now we want to establish (4.12). The quickest way to do it is to use some results of Hajłasz and Koskela. We say that (u, g) forms a Poincaré pair if u is in $L^1_{loc}(\mathbb{R}^n, w)$, g is positive and measurable and for any ball $B \subset \mathbb{R}^n$ of radius r, we have

$$m(B)^{-1} \int_B |u(z) - u_B| dm(z) \leq Crm(B)^{-1} \int_B g\, dm(z). \tag{4.24}$$

In this context, Theorem 5.1 (and Corollary 9.8) in [**HaK2**] states that the Poincaré inequality (4.24) can be improved into a Sobolev-Poincaré inequality.

More precisely, if s is such that, for any ball B_0 of radius r_0, any $x \in B_0$ and any $r \leq r_0$,

$$\frac{m(B(x,r))}{m(B_0)} \geq C^{-1}\left(\frac{r}{r_0}\right)^s \tag{4.25}$$

then (4.24) implies for any $1 < q < s$

$$\left(m(B)^{-1}\int_B |u(z) - u_B|^{q^*} dm(z)\right)^{\frac{1}{q^*}} \leq Cr\left(m(B)^{-1}\int_B g^q dm(z)\right)^{\frac{1}{q}} \tag{4.26}$$

where $q^* = \frac{qs}{s-q}$ and B is a ball of radius r. Combined with Hölder's inequality, we get

$$\left(m(B)^{-1}\int_B |u(z) - u_B|^p dm(z)\right)^{\frac{1}{p}} \leq Cr\left(m(B)^{-1}\int_B g^2 dm(z)\right)^{\frac{1}{2}} \tag{4.27}$$

for any $p \in [1, 2s/(s-2)]$ if $s > 2$ or any $p < +\infty$ if $s \leq 2$.

We will use the result of Hajłasz and Koskela with $g = |\nabla u|$. We need to check the assumptions of their result. The bound (4.24) is exactly (4.17) and we already proved it. The second and last thing we need to verify is that (4.25) holds with $s = n$. This fact is an easy consequence of (2.8). Indeed, if B_0 is a ball of radius r_0, $x \in B_0$ and $r \leq r_0$

$$\frac{m(B(x,r))}{m(B_0)} \geq \frac{m(B(x,r))}{m(B(x,2r_0))}. \tag{4.28}$$

Yet, (2.8) implies that $\frac{m(B(x,r))}{m(B(x,2r_0))}$ is bounded from below by $C^{-1}(\frac{r}{2r_0})^n$, that is $C^{-1}(\frac{r}{r_0})^n$. Then

$$\frac{m(B(x,r))}{m(B_0)} \geq C^{-1}\left(\frac{r}{r_0}\right)^n, \tag{4.29}$$

which is the desired conclusion. We deduce that (4.27) holds with $g = |\nabla u|$ and for any $p \in [1, \frac{2n}{n-2}]$ ($1 \leq p < +\infty$ if $n = 2$), which is exactly (4.12).

To finish to prove the lemma, it remains to establish (4.13). Let $B = B(x,r)$ be a ball centered on Γ. However, since $x \in \Gamma$, (2.2) entails that $m(B)$ is equivalent to r^{d+1}. Thus, thanks to (4.12) and Lemma 4.1,

$$\begin{aligned}\left(r^{-d-1}\int_B |u(z)|^p dm(z)\right)^{\frac{1}{p}} &\leq C\left(m(B)^{-1}\int_B \left|u(z) - \fint_B u\,dy\right|^p dm(z)\right)^{\frac{1}{p}} + C\fint_B |u(z)|dz \\ &\leq Cr\left(r^{-d-1}\int_B |\nabla u|^2 dm(z)\right)^{\frac{1}{2}},\end{aligned} \tag{4.30}$$

which proves Lemma 4.2. □

REMARK 4.3. If $B \subset \mathbb{R}^n$ and $u \in W$ is supported in B, then for any $p \in [1, 2n/(n-2)]$ (or $p \in [1, +\infty)$ if $n = 2$), there holds

$$\left\{\frac{1}{m(B)}\int_B |u(y)|^p\, w(y)dy\right\}^{1/p} \leq Cr\left\{\frac{1}{m(B)}\int_B |\nabla u(y)|^2 w(y)dy\right\}^{1/2}. \tag{4.31}$$

That is, we can choose $u_B = 0$ in (4.12).

To prove (4.31), the main idea is that we can replace in (4.12) the quantity $u_B = m(B)^{-1}\int_B u$ by the average $u_{\bar{B}}$, where $\bar{B}$ is a ball near B. We choose for $\bar{B}$ a ball with same radius as B and contained in $3B \setminus B$, because this way $u_{\bar{B}} = 0$ since u is supported in B. Then

$$\begin{aligned} &\left\{\frac{1}{m(3B)}\int_{3B}|u(y)|^p w(y)dy\right\}^{1/p} \\ &\quad= \left\{\frac{1}{m(3B)}\int_{3B}|u(y)-u_{\bar{B}}|^p w(y)dy\right\}^{1/p} \\ &\quad\le \left\{\frac{1}{m(3B)}\int_{3B}|u(y)-u_{3B}|^p w(y)dy\right\}^{1/p} + |u_{3B}-u_{\bar{B}}| \qquad (4.32) \end{aligned}$$

Yet, using Jensen's inequality and then Hölder's inequality, $|u_{3B}-u_{\bar{B}}|$ is bounded by $\left\{\frac{1}{m(\bar{B})}\int_{3B}|u(y)-u_{3B}|^p w(y)dy\right\}^{1/p}$. If we use in addition the doubling property given by (4.29), we get that $|u_{3B}-u_{\bar{B}}|$ is bounded by $\left\{\frac{1}{m(3B)}\int_{3B}|u(y)-u_{3B}|^p w(y)dy\right\}^{1/p}$, that is,

$$\left\{\frac{1}{m(3B)}\int_{3B}|u(y)|^p w(y)dy\right\}^{1/p} \le \left\{\frac{1}{m(3B)}\int_{3B}|u(y)-u_{3B}|^p w(y)dy\right\}^{1/p}. \qquad (4.33)$$

We conclude thanks to (4.12) and the doubling property (2.8).

CHAPTER 5

Completeness and Density of Smooth Functions

In later chapters we shall work with various dense classes. We prepare the job in this section, with a little bit of work on function spaces and approximation arguments. Most results in this section are basically unsurprising, except perhaps the fact that when $d \leq 1$, the test functions are dense in W (with no decay condition at infinity).

Let $\dot{W}$ be the factor space $W/\mathbb{R}$, equipped with the norm $\|\cdot\|_W$. The elements of $\dot{W}$ are classes $\dot{u} = \{u + c\}_{c \in \mathbb{R}}$, where $u \in W$.

LEMMA 5.1. *The space $\dot{W}$ is complete. In particular, if a sequence of elements of W, $\{v_k\}_{k=1}^{\infty}$, and $u \in W$ are such that $\|v_k - u\|_W \to 0$ as $k \to \infty$, then there exist constants $c_k \in \mathbb{R}$ such that $v_k - c_k \to u$ in $L^1_{loc}(\mathbb{R}^n)$.*

PROOF. Let $(\dot{u}_k)_{k \in \mathbb{N}}$ be a Cauchy sequence in $\dot{W}$. We need to show that

(i) for every sequence $(v_k)_{k \in \mathbb{N}}$ in W, with $v_k \in \dot{u}_k$ for $k \in \mathbb{N}$, there exists $u \in W$ and $(c_k)_{k \in \mathbb{N}}$ such that $v_k - c_k \to u$ in $L^1_{loc}(\mathbb{R}^n)$ and

$$\lim_{k \to \infty} \|v_k - u\|_W = 0; \tag{5.1}$$

(ii) if u and u' are such that there exist $(v_k)_{k \in \mathbb{N}}$ and $(v'_k)_{k \in \mathbb{N}}$ such that $v_k, v'_k \in \dot{u}_k$ for all $k \in \mathbb{N}$ and

$$\lim_{k \to \infty} \|v_k - u\|_W = \lim_{k \to \infty} \|v'_k - u'\|_W = 0, \tag{5.2}$$

then $\dot{u}' = \dot{u}$.

First assume that (i) is true and let us prove (ii). Let u, u', $(v_k)_{k \in \mathbb{N}}$ and $(v'_k)_{k \in \mathbb{N}}$ be such that $v_k, v'_k \in \dot{u}_k$ for any $k \in \mathbb{N}$ and (5.2) holds. Then the sequence $(\nabla v_k - \nabla v'_k)_{k \in \mathbb{N}}$ converges in $L^2(\Omega, w)$ to $\nabla(u - u')$ on one hand and is constant equal to 0 on the other hand. Thus $\nabla(u - u') = 0$ and u and u' differ only by a constant, hence $\dot{u}' = \dot{u}$.

Now we prove (i). By translation invariance, we may assume that $0 \in \Gamma$. Let the $v_k \in \dot{u}_k$ be given, and choose $c_k = \fint_{B(0,1)} v_k$. We want to show that $v_k - c_k$ converges in $L^1_{loc}(\mathbb{R}^n)$.

Set $B_j = B(0, 2^j)$ for $j \geq 0$; let us check that for $f \in W$ and $j \geq 0$,

$$\fint_{B_j} \left| f(x) - \fint_{B_0} f(y)\, dy \right| dx \leq C 2^{(n+1)j} \|f\|_W. \tag{5.3}$$

Set $m_j = \fint_{B_j} f\,dy$; observe that

$$
\begin{aligned}
\fint_{B_j} |f - m_j|\,dx &\leq \frac{1}{m(B_j)} \int_{B_j} |f(x) - m_j| w(x) dx \\
&\leq C2^j m(B_j)^{-1/2} \Big\{ \int_{B_j} |\nabla f(y)|^2 w(y) dy \Big\}^{1/2} \qquad (5.4)\\
&\leq C2^j m(B_j)^{-1/2} ||f||_W \leq C2^j ||f||_W
\end{aligned}
$$

by (2.13), the Poincaré inequality (4.12) with $p = 1$, and a brutal estimate using (2.2), our assumption that $0 \in \Gamma$, and the fact that $B_j \supset B_0$. In addition,

$$
\begin{aligned}
|m_0 - m_j| = \Big| \fint_{B_0} f(x)\,dx - m_j \Big| &\leq \fint_{B_0} |f(x) - m_j|\,dx \\
&\leq 2^{jn} \fint_{B_j} |f(x) - m_j|\,dx \leq C2^{(n+1)j} \|f\|_W
\end{aligned} \qquad (5.5)
$$

by (5.4). Finally

(5.6)

$$
\begin{aligned}
\fint_{B_j} \Big| f(x) - \fint_{B_0} f(y) dy \Big| dx = \fint_{B_j} \Big| f(x) - m_0 \Big| dx &\leq \fint_{B_j} |f(x) - m_j| dx + |m_0 - m_j| \\
&\leq C2^{(n+1)j} ||f||_W,
\end{aligned}
$$

as needed for (5.3).

Return to the convergence of v_k. Recall that $c_k = \fint_{B_1} v_k$. By (5.3) with $f = v_k - c_k - v_l + c_l$ (so that $m_0 = 0$), $v_k - c_k$ is a Cauchy sequence in $L^1_{loc}(B_j)$ for each $j \geq 0$, hence there exists $u^j \in L^1(B_j)$ such that $v_k - c_k$ converges to u^j. By uniqueness of the limit, we have that for $1 \leq j \leq j_0$,

$$u^{j_0} = u^j \quad \text{a.e. in } B_j \qquad (5.7)$$

and thus we can define a function u on $\mathbb{R}^n$ as $u(x) = u^j(x)$ if $x \in B_j$. By construction $u \in L^1_{loc}(\mathbb{R}^n)$ and $v_k - c_k \to u$ in $L^1_{loc}(\mathbb{R}^n)$.

It remains to show that u is actually in W and $v_k \to u$ in W. First, since $L^2(\Omega, w)$ is complete, there exists V such that ∇v_k converges to V in $L^2(\Omega, w)$. Then observe that for $\varphi \in C_0^\infty(B_j \setminus \Gamma, \mathbb{R}^n)$,

$$\int_{B_j} V \cdot \varphi = \lim_{k \to \infty} \int_{B_j} \nabla v_k \cdot \varphi = - \lim_{k \to \infty} \int_{B_j} (v_k - c_k) \operatorname{div} \varphi = - \int_{B_j} u^j \operatorname{div} \varphi.$$

Hence by definition of a weak derivative,

$$\nabla u = \nabla u^j = V \quad \text{a.e. in } B_j.$$

Since the result holds for any $j \geq 1$,

$$\nabla u = V \quad \text{a.e. in } \mathbb{R}^n,$$

that is, by construction of V, $u \in W$ and $\|v_k - u\|_W$ converges to 0. □

LEMMA 5.2. *The space*

$$W_0 = \{u \in W\,;\, Tu = 0\}, \qquad (5.8)$$

equipped with the scalar product $\langle u, v \rangle_W := \int_\Omega \nabla u(z) \cdot \nabla v(z)\, dm(z)$ *(and the norm* $\|.\|_W$*) is a Hilbert space.*

Moreover, for any ball B centered on Γ, the set

$$W_{0,B} = \{u \in W\,;\, Tu = 0 \ \ \mathcal{H}^d\text{-almost everywhere on } \Gamma \cap B\}, \tag{5.9}$$

equipped with the scalar product $\langle .,.\rangle_W$, is also a Hilbert space.

PROOF. Observe that W_0 and $W_{0,B}$ are no longer spaces of functions defined modulo an additive constant. That is, if $f \in W_0$ (or $W_{0,B}$) is a constant c, then $c = 0$ because (3.10) says that $Tu = c$ almost everywhere on Γ. Thus $\|.\|_W$ is really a norm on W_0 and $W_{0,B}$, and we only need to prove that these spaces are complete. We first prove this for $W_{0,B}$; the case of W_0 will be easy deal with afterwards.

Let B be a ball centered on Γ, and consider $W_{0,B}$. By translation and dilation invariance of the result, we can assume that $B = B(0,1)$.

Let $(v_k)_{k\in\mathbb{N}}$ be a Cauchy sequence of functions in $W_{0,B}$. We want first to show that v_k has a limit in $L^1_{loc}(\mathbb{R}^n)$ and W. We use Lemma 5.1 and so there exists $\bar{u} \in W$ and $c_k \in \mathbb{R}$ such that

$$\|v_k - \bar{u}\|_W \to 0 \tag{5.10}$$

and

$$v_k - c_k \to \bar{u} \quad \text{in } L^1_{loc}(\mathbb{R}^n). \tag{5.11}$$

By looking at the proof of Lemma 5.1, we can take $c_k = \fint_B v_k$. Let us prove that (c_k) is a Cauchy sequence in $\mathbb{R}$. We have for any $k, l \geq 0$

$$|c_k - c_l| \leq \fint_B |v_k(z) - v_l(z)|\,dz \leq Cm(B)^{-1}\int_B |v_k(z) - v_l(z)|dm(z) \tag{5.12}$$

with (2.13). Since $T(v_k - v_l) = 0$ on B and since we assumed that B is the unit ball, Lemma 4.2 entails

$$|c_k - c_l| \leq C\|v_k - v_l\|_W. \tag{5.13}$$

Since $(v_k)_{k\in\mathbb{N}}$ is a Cauchy sequence in W, $(c_k)_{k\in\mathbb{N}}$ is a Cauchy sequence in $\mathbb{R}$ and thus converges to some value $c \in \mathbb{R}$. Set $u = \bar{u} - c$. We deduce from (5.11) that

$$v_k \to u \quad \text{in } L^1_{loc}(\mathbb{R}^n), \tag{5.14}$$

and since u and $\bar{u}$ differ only from a constant, (5.10) can be rewritten as

$$\|v_k - u\|_W \to 0. \tag{5.15}$$

We still need to show that $u \in W_{0,B}$, i.e., that $Tu = 0$ a.e. on B. We will actually prove something a bit stronger. We claim that if $u, v_k \in W$, then the convergence of v_k to u in both W and $L^1_{loc}(\mathbb{R}^n)$ implies the convergence of the traces $Tv_k \to Tu$ in $L^1_{loc}(\Gamma,\sigma)$. That is,

$$v_k \to u \text{ in } W \text{ and in } L^1_{loc}(\mathbb{R}^n) \implies Tv_k \to Tu \text{ in } L^1_{loc}(\Gamma,\sigma). \tag{5.16}$$

Recall that by (3.26), $Tf \in L^1_{loc}(\Gamma,\sigma)$ whenever $f \in W$. Our result, that is $Tu = 0$ a.e. on B, follows easily from the claim: we already established that $v_k \to u$ in W and in $L^1_{loc}(\mathbb{R}^n)$ and thus (5.16) gives that $\|Tu\|_{L^1(B,\sigma)} = \lim_{k\to\infty}\|Tv_k\|_{L^1(B,\sigma)} = 0$, i.e., that $Tu = 0$ σ-a.e. in B.

We turn to the proof of (5.16). Since T is linear, we may subtract u, and assume that v_k tends to 0 and $u = 0$. Let us use the notation of Theorem 3.4, and set $g^k = Tv_k$ and $g^k_r(x) = \fint_{B(x,r)} v_k(z)\,dz$. Since $\|v_k\|_W$ tends to 0, we may assume without loss of generality that $\|v_k\|_W \leq 1$ for $k \in \mathbb{N}$. We want to prove that for every ball $\widetilde{B} \subset \mathbb{R}^n$ centered on Γ and every $\epsilon > 0$, we can find k_0 such that

$$\|g^k\|_{L^1(\widetilde{B},\sigma)} \leq \epsilon \quad \text{for } k \geq k_0. \tag{5.17}$$

We may also assume that the radius of $\widetilde{B}$ is larger than 1 (as it makes (5.17) harder to prove).

Fix $\widetilde{B}$ and ε as above, and $\alpha \in (0, 1/2)$, and observe that for $r \in (0,1)$,

$$\begin{aligned} \int_{\widetilde{B}} |g^k|\, d\sigma &\leq \int_{\widetilde{B}} |g^k - g_r^k| d\sigma + \int_{\widetilde{B}} |g_r^k| d\sigma \\ &\leq C(\widetilde{B})\|g^k - g_r^k\|_{L^2(\sigma)} + \int_{x\in\widetilde{B}} \fint_{y\in B(x,r)} |v_k(y)| dy d\sigma(x) \\ &\leq C(\widetilde{B},\alpha) r^{2\alpha}\|v_k\|_W + Cr^{d-n}\int_{2\widetilde{B}} |v_k(y)|dy, \end{aligned} \tag{5.18}$$

where for the last line we used (3.24), Fubini, and the condition (1.1) on Γ. Recall that $\|v_k\|_W \leq 1$; we choose r so small that $C(\widetilde{B},\alpha)r^{2\alpha}\|u\|_W \leq \epsilon/2$, and since by assumption v_k tends to 0 in L^1_{loc}, we can find k_0 such that $Cr^{d-n}\int_{2\widetilde{B}} |v_k(y)|dy \leq \varepsilon/2$ for $k \geq k_0$, as needed for (5.17).

This completes the proof of (5.16), and we have seen that the completeness of $W_{0,B}$ follows. Since W_0 is merely an intersection of spaces $W_{0,B}$, it is complete as well, and Lemma 5.2 follows. □

LEMMA 5.3. *Choose a non-negative function $\rho \in C_0^\infty(\mathbb{R}^n)$ such that $\int \rho = 1$ and ρ is supported in $\overline{B(0,1)}$. Furthermore let ρ be radial and nonincreasing, i.e. $\rho(x) = \rho(y) \geq \rho(z)$ if $|x| = |y| \leq |z|$. Define ρ_ϵ, for $\epsilon > 0$, by $\rho_\epsilon(x) = \epsilon^{-n}\rho(\epsilon^{-1}x)$. For every $u \in W$, we have:*

(i) *$\rho_\epsilon * u \in C^\infty(\mathbb{R}^n)$ for every $\epsilon > 0$;*
(ii) *If $x \in \mathbb{R}^n$ is a Lebesgue point of u, then $\rho_\epsilon * u(x) \to u(x)$ as $\epsilon \to 0$; in particular, $\rho_\epsilon * u \to u$ a.e. in $\mathbb{R}^n$;*
(iii) *$\nabla(\rho_\epsilon * u) = \rho_\epsilon * \nabla u$ for $\varepsilon > 0$;*
(iv) *$\lim_{\epsilon\to 0} \|\rho_\epsilon * u - u\|_W = 0$;*
(v) *$\rho_\epsilon * u \to u$ in $L^1_{loc}(\mathbb{R}^n)$.*

PROOF. Recall that $W \subset L^1_{loc}(\mathbb{R}^n)$ (see Lemma 3.2). Thus conclusions (i) and (ii) are classical and can be found as Theorem 1.12 in [**MZ**].

Let $u \in W$ and write u_ϵ for $\rho_\epsilon * u$. We have seen that $u_\epsilon \in C^\infty(\mathbb{R}^n)$, so ∇u_ϵ is defined on $\mathbb{R}^n$. One would like to say that $\nabla u_\epsilon = \rho_\epsilon * \nabla u$, i.e. point (iii). Here ∇u_ϵ is the classical gradient of u_ϵ on $\mathbb{R}^n$, thus *a fortiori* also the distributional gradient on $\mathbb{R}^n$ of u_ϵ. That is, for any $\varphi \in C_0^\infty(\mathbb{R}^n, \mathbb{R}^n)$, there holds

(5.19)
$$\begin{aligned} \int_{\mathbb{R}^n} \nabla u_\epsilon \cdot \varphi = -\int_{\mathbb{R}^n} u_\epsilon(x) \operatorname{div}\varphi(x) dx = -\int_{\mathbb{R}^n}\int_{\mathbb{R}^n} \rho_\epsilon(y) u(x-y) \operatorname{div}\varphi(x) dy\, dx \\ = \int_{B(0,\epsilon)} \rho_\epsilon(y) \left(-\int_{\mathbb{R}^n} u(z) \operatorname{div}\varphi(z+y) dz \right) dy. \end{aligned}$$

The function φ lies in $C_0^\infty(\mathbb{R}^n, \mathbb{R}^n)$, and so does, for any $y \in \mathbb{R}^n$, the function $z \mapsto \varphi(z+y)$. Recall that ∇u is the distributional derivative of u on Ω but yet also the distributional derivative of u on $\mathbb{R}^n$ (see Lemma 3.2). Therefore

$$\begin{aligned}\int_{\mathbb{R}^n} \nabla u_\epsilon \cdot \varphi &= \int_{B(0,\epsilon)} \rho_\epsilon(y) \int_{\mathbb{R}^n} \nabla u(z) \cdot \varphi(z+y) dz\, dy \\ &= \int_{\mathbb{R}^n} \rho_\epsilon(y) \int_{\mathbb{R}^n} \nabla u(x-y) \cdot \varphi(x) dx\, dy = \int_{\mathbb{R}^n} (\rho_\epsilon * \nabla u) \cdot \varphi,\end{aligned} \tag{5.20}$$

which gives (iii).

From there, our point (iv), that is the convergence of $\rho_\epsilon * u$ to u in W, can be deduced with, for instance, [**Kil**, Lemma 1.5]. The latter states that, under our assumptions on ρ, the convergence $\rho_\epsilon * g \to g$ holds in $L^2(\mathbb{R}^n, w)$ whenever $g \in L^2(\mathbb{R}^n, w)$ and w is in the Muckenhoupt class $\mathcal{A}_2$ (we already proved this fact, see Lemma 2.5). Note that Kilpelai's result is basically a consequence of a result from Muckenhoupt about the boundedness of the (unweighted) Hardy-Littlewood maximal function in weighted L^p.

Finally we need to prove (v). Just notice that $u \in L^1_{loc}(\mathbb{R}^n)$, and apply the standard proof of the fact that $\rho_\epsilon * u \to u$ in L^1 for $f \in L^1$. The lemma follows $\square$

LEMMA 5.4. *Let $u \in W$ and $\varphi \in C_0^\infty(\mathbb{R}^n)$. Then $u\varphi \in W$ and for any point $x \in \Gamma$ satisfying* (3.11)

$$T(u\varphi)(x) = \varphi(x) Tu(x). \tag{5.21}$$

PROOF. The function u lies in $L^1_{loc}(\mathbb{R}^n)$ and thus defines a distribution on $\mathbb{R}^n$ (see Lemma 3.2). Multiplication by smooth functions and (distributional) derivatives are always defined for distributions and, in the sense of distribution, $\nabla(u\varphi) = \varphi \nabla u + u \nabla \varphi$. Let $B \subset \mathbb{R}^n$ be a big ball such that $\operatorname{supp} \varphi \subset B$. Then

$$\begin{aligned}\|u\varphi\|_W &\leq \|\varphi\|_\infty \|\nabla u\|_{L^2(\Omega,w)} + \|\nabla \varphi\|_\infty \Big\| u - \fint_B u \Big\|_{L^2(B,w)} + \|\nabla\varphi\|_\infty \Big\| \fint_B u \Big\|_{L^2(B,w)} \\ &\leq \|\varphi\|_\infty \|\nabla u\|_{L^2(\Omega,w)} + C_B \|\nabla \varphi\|_\infty \|u\|_W + C_B \|\nabla \varphi\|_\infty \|u\|_{L^1(B)} < +\infty\end{aligned} \tag{5.22}$$

by the Poincaré inequality (4.12). We deduce $u\varphi \in W$.

Let take a Lebesgue point x satisfying (3.11). We have

$$\begin{aligned}&\fint_{B(x,r)} |u(z)\varphi(z) - \varphi(x) Tu(x)|\, dz \\ &\qquad \leq \fint_{B(x,r)} |u(z) - Tu(x)||\varphi(z)|\, dz + |Tu(x)| \fint_{B(x,r)} |\varphi(z) - \varphi(x)|\, dz \\ &\qquad \leq \|\varphi\|_\infty \fint_{B(x,r)} |u(z) - Tu(x)|\, dz + |Tu(x)| \fint_{B(x,r)} |\varphi(z) - \varphi(x)|\, dz.\end{aligned}$$

The first term of the right-hand side converges to 0 because x is a Lebesgue point. The second term in the right-hand side converges to 0 because φ is continuous. The equality (5.21) follows. $\square$

Let F be a closed set in $\mathbb{R}^n$ and $E = \mathbb{R}^n \setminus F$. In the sequel, we let

$$C_c^\infty(E) = \{f \in C^\infty(E), \exists \epsilon > 0 \text{ such that } f(x) = 0 \text{ whenever } \operatorname{dist}(x, F) \leq \varepsilon\} \tag{5.23}$$

denote the set of functions in $C^\infty(E)$ that equal 0 in a neighborhood of F. Furthermore, we use the notation $C_0^\infty(E)$ for the set of functions that are compactly supported in E, that is

$$C_0^\infty(E) = \{f \in C_c^\infty(E), \exists R > 0 : \text{supp} f \subset B(0,R)\}. \tag{5.24}$$

LEMMA 5.5. *The completion of $C_0^\infty(\Omega)$ for the norm $\|.\|_W$ is the set*

$$W_0 = \{u \in W \,;\, Tu = 0\} \tag{5.25}$$

of (5.8). *Moreover, if $u \in W_0$ is supported in a compact subset of the open ball $B \subset \mathbb{R}^n$, then u can be approximated in the W-norm by functions of $C_0^\infty(B \setminus \Gamma)$.*

PROOF. The proof of this result will use two main steps, where

(i) we use cut-off functions φ_r to approach any function $u \in W_0$ by functions in W that equal 0 on a neighborhood of Γ;
(ii) we use cut-off functions ϕ_R to approach any function $u \in W_0$ by functions in W that are compactly supported in $\mathbb{R}^n$.

Part (i): For $r > 0$ small, we choose a smooth function φ_r such that $\varphi(x) = 0$ when $\delta(x) \leq r$, $\varphi(x) = 1$ when $\delta(x) \geq 2r$, $0 \leq \varphi \leq 1$ everywhere, and $|\nabla\varphi(x)| \leq 10r^{-1}$ everywhere.

Let $u \in W_0$ be given. We want to show that for r small, $\varphi_r u$ lies in W and

$$\lim_{r\to 0} \|u - \varphi_r u\|_W^2 = 0. \tag{5.26}$$

Notice that $\varphi_r u \in L^1_{loc}(\Omega)$, just like u, and its distribution gradient on Ω is locally in L^2 and given by

$$\nabla(\varphi_r u)(x) = \varphi_r(x)\nabla u(x) + u(x)\nabla\varphi_r(x). \tag{5.27}$$

So we just need to show that

$$\begin{aligned} &\lim_{r\to 0} \int |\nabla(\varphi_r u)(x) - \nabla u(x)|^2 w(x)dx \\ &= \lim_{r\to 0} \int |u(x)\nabla\varphi_r(x) + (1-\varphi_r(x))\nabla u(x)|^2 w(x)dx = 0. \end{aligned} \tag{5.28}$$

Now $\int |\nabla u(x)|^2 w(x)dx = \|u\|_W^2 < +\infty$, so $\int |(1-\varphi_r)\nabla u(x)|^2 w(x)dx$ tends to 0, by the dominated convergence theorem, and it is enough to show that

$$\lim_{r\to 0} \int |u(x)\nabla\varphi_r(x)|^2 w(x)dx = 0. \tag{5.29}$$

Cover Γ with balls B_j, $j \in J$, of radius r, centered on Γ, and such that the $3B_j$ have bounded overlap, and notice that the region where $\nabla\varphi_r \neq 0$ is contained in $\cup_{j\in J} 3B_j$. In addition, if $x \in 3B_j$ is such that $\nabla\varphi_r \neq 0$, then $|\nabla\varphi_r(x)| \leq 10r^{-1}$, so that

(5.30)
$$\int_{3B_j} |u(x)\nabla\varphi_r(x)|^2 w(x)dx \leq 100r^{-2} \int_{3B_j} |u(x)|^2 w(x)dx \leq C \int_{3B_j} |\nabla u(x)|^2 w(x)dx,$$

where the last part comes from (4.13), applied with $p = 2$ and justified by the fact that $Tu = 0$ on the whole Γ. We may now sum over j. Denote by A_r the union of the $3B_j$; then

$$\begin{aligned}
&\int_\Omega |u(x)\nabla\varphi_r(x)|^2 w(x)dx \\
&\qquad \le \sum_{j\in J}\int_{3B_j} |u(x)\nabla\varphi_r(x)|^2 w(x)dx \le C\sum_{j\in J}\int_{3B_j} |\nabla u(x)|^2 w(x)dx \\
&\qquad \le C\int_{A_r} |\nabla u(x)|^2 w(x)dx \qquad (5.31)
\end{aligned}$$

because the $3B_j$ have bounded overlap. The right-hand side of (5.31) tends to 0, because $\int_\Omega |\nabla u(x)|^2 w(x)dx = \|u\|_W^2 < +\infty$ and by the dominated convergence theorem. The claim (5.29) follows, and so does (5.26). This completes Part (i).

Part (ii). By translation invariance, we may assume that $0 \in \Gamma$. Let R be a big radius; we want to define a cut-off function ϕ_R.

If we used the classical cut-off function built as $\bar\phi_R = \bar\phi\left(\frac{x}{R}\right)$ with $\bar\phi$ supported in $B(0,1)$, the convergence would work with the help of Poincaré's inequality on annuli. But since we did not prove this inequality, we will proceed differently and use the 'better' cut-off functions defined as follows.

Set $\phi_R(x) = \phi\left(\frac{\ln|x|}{\ln R}\right)$, where ϕ is a smooth function defined on $[0,+\infty)$, supported in $[0,1]$ and such that $\phi \equiv 1$ on $[0,1/2]$. In particular, one can see that $\nabla\phi_R(x) \le \frac{C}{\ln R}\frac{1}{|x|}$ and that $\nabla\phi_R$ is supported in $\{x \in \mathbb{R}^n,\ \sqrt{R} \le |x| \le R\}$. We take $\widehat{u} := \phi_R u$ and we want to show that $\widehat{u} \in W$ and $\|u - \widehat{u}\|_W$ is small. Notice that $\widehat{u} \in W_0$, by Lemma 5.4, and in addition $\widehat{u}$ is supported in $B(0,R)$. We want to show that

$$\lim_{R\to+\infty} \|\widehat{u} - u\|_W^2 = 0. \qquad (5.32)$$

But $\widehat{u} \in L^1_{loc}(\Omega)$, just like u, and its distribution gradient on Ω is locally in L^1 and given by

$$\nabla\widehat{u}(x) = \phi_R(x)\nabla u(x) + u(x)\nabla\phi_R(x). \qquad (5.33)$$

Hence

$$\begin{aligned}
\|\widehat{u} - u\|_W^2 &= \int |\nabla\widehat{u}(x) - \nabla u(x)|^2 w(x)dx \\
&= \int |u(x)\nabla\phi_R(x) + (1-\phi_R(x))\nabla u(x)|^2 w(x)dx. \qquad (5.34)
\end{aligned}$$

Now $\int |\nabla u(x)|^2 w(x)dx = \|u\|_W^2 < +\infty$, so $\int |(1-\phi_R)\nabla u(x)|^2 w(x)dx$ tends to 0, by the dominated convergence theorem, and it is enough to show that

$$\lim_{R\to+\infty} \int |u(x)\nabla\phi_R(x)|^2 w(x)dx = 0. \qquad (5.35)$$

Let C_j be the annulus $\{x \in \mathbb{R}^n,\ 2^j < |x| \le 2^{j+1}\}$. The bounds on $\nabla\phi_R$ yield

$$\int |u(x)\nabla\phi_R(x)|^2 w(x)dx \le \frac{C}{(\ln R)^2}\sum_{j=0}^{1+\log_2 R} 2^{-2j}\int_{C_j} |u(x)|^2 w(x)dx. \qquad (5.36)$$

The integral on the annulus C_j is smaller than the integral in the ball $B(0, 2^{j+1})$. Since $u \in W_0$ and $0 \in \Gamma$, (4.13) yields

$$\begin{aligned}\int |u(x)\nabla\phi_R(x)|^2 w(x)dx &\leq \frac{C}{(\ln R)^2} \sum_{j=0}^{1+\log_2 R} \int_{B(0,2^{j+1})} |\nabla u(x)|^2 w(x)dx \\ &\leq \frac{C}{(\ln R)^2} \|u\|_W^2 \sum_{j=0}^{1+\log_2 R} 1 \leq \frac{C}{|\ln R|} \|u\|_W.\end{aligned} \tag{5.37}$$

Thus $\int |u(x)\nabla\phi_R(x)|^2 w(x)dx$ converges to 0 as R goes to $+\infty$, which proves (5.35) and ends Part (ii).

We are now ready to prove the lemma. If $u \in W_0$ and $\varepsilon > 0$ is given, we can find R such that $\|\phi_R u - u\|_W^2 \leq \varepsilon$ (by (5.32)). Notice that $\phi_R u \in W_0$, by Lemma 5.4, and now we can find r such that $\|\varphi_r \phi_R u - \phi_R u\|_W^2 \leq \varepsilon$ (by (5.26)). In turn $\varphi_r \phi_R u$ is compactly supported away from Γ, and we may now use Lemma 5.3 to approximate it with smooth functions with compact support in Ω. It follows that W_0 is included in the completion of $C_0^\infty(\Omega)$. Since W_0 is complete (see Lemma 5.2), the reverse inclusion is immediate.

For the second part of the lemma, we are given $u \in W_0$ with a compact support inside B, we can use Part (i) to approximate it by some $\varphi_r u$ with a compact support inside B. A convolution as in Lemma 5.3 then makes it smooth without destroying the support property; Lemma 5.5 follows. □

REMARK 5.6. We don't know how to prove exactly the same result for the spaces $W_{0,B}$ of (5.9). However, we have the following weaker result. Let $B \subset \mathbb{R}^n$ be a ball and $B_{\frac{1}{2}}$ denotes the ball with same center as B but half its radius. For any function $u \in W_{0,B}$, there exists a sequence $(u_k)_{k\in\mathbb{N}}$ of functions in $C_c^\infty(\mathbb{R}^n \setminus \overline{B_{\frac{1}{2}} \cap \Gamma})$ such that $\|u_k - u\|_W$ converges to 0.

Indeed, take $\eta \in C_0^\infty(B)$ such that $\eta = 1$ on $B_{\frac{3}{4}}$. Write $u = \eta u + (1-\eta)u$; it is enough to prove that both ηu and $(1-\eta)u$ can be approximated by functions in $C_c^\infty(\mathbb{R}^n \setminus \overline{B_{\frac{1}{2}} \cap \Gamma})$. Notice first that $\eta u \in W_0$ and thus can be approximated by functions in $C_0^\infty(\Omega) \subset C_c^\infty(\mathbb{R}^n \setminus \overline{B_{\frac{1}{2}} \cap \Gamma})$, according to Lemma 5.5. Besides, $(1-\eta)u$ is supported outside of $B_{\frac{3}{4}}$ and thus, if ϵ is smaller than a quarter of the radius of B, then the functions $\rho_\epsilon * [(1-\eta)u]$ are in $C_c^\infty(\mathbb{R}^n \setminus \overline{B_{\frac{1}{2}}}) \subset C_c^\infty(\mathbb{R}^n \setminus \overline{B_{\frac{1}{2}} \cap \Gamma})$. Lemma 5.3 gives then that the family $\rho_\epsilon * [(1-\eta)u]$ approaches $(1-\eta)u$ as ϵ goes to 0.

Next we worry about the completion of $C_0^\infty(\mathbb{R}^n)$ for the norm $\|.\|_W$. We start with the case when $d > 1$; when $0 < d \leq 1$, things are a little different and they will be discussed in Lemma 5.11.

LEMMA 5.7. *Let $d > 1$. Choose $x_0 \in \Gamma$ and write B_j for $B(x_0, 2^j)$. Then for any $u \in W$*

$$u^0 := \lim_{j\to+\infty} \fint_{B_j} u(z)\, dz \textit{ exists and is finite.} \tag{5.38}$$

The completion of $C_0^\infty(\mathbb{R}^n)$ for the norm $\|.\|_W$ can be identified to a subspace of $L^1_{loc}(\mathbb{R}^n)$, which is

$$W^0 = \{u \in W,\, u^0 = 0\}. \tag{5.39}$$

REMARK 5.8. Since $C_0^\infty(\Omega) \subset C_0^\infty(\mathbb{R}^n)$, Lemmata 5.5 and 5.7 imply that $W_0 \subset W^0$. In particular, we get that

$$\lim_{j\to+\infty} \fint_{B_j} u(z)\, dz = 0 \quad \text{for } u \in W_0. \tag{5.40}$$

REMARK 5.9. Since the completion of $C_0^\infty(\mathbb{R}^n)$ doesn't depend on our choice of x_0, the value u^0 doesn't depend on x_0 either. Similarly, with a small modification in the proof, we could replace (2^j) with any other sequence that tends to $+\infty$.

REMARK 5.10. The lemma immediately implies the following result: for any $u \in W$, $u - u^0 \in W^0$ and thus can be approximated in $L^1_{loc}(\mathbb{R}^n)$ and in the W-norm by function in $C_0^\infty(\mathbb{R}^n)$.

PROOF. Let $d > 1$ and choose $u \in W$. Let us first prove that u^0 is well defined. By translation invariance, we can choose $x_0 = 0$, that is $B_j = B(0, 2^j)$. For $j \in \mathbb{N}$, set $u_j = \fint_{B_j} f(z)\, dz$ and $V_j = \int_{B_j} dm$. The bounds (2.2) give that V_j is equivalent to $2^{j(1+d)}$ and (2.14) gives that for any $z \in B_j$, $\frac{V_j}{|B_j|} \leq Cw(z)$. Then by Lemma 4.2

$$\begin{aligned} |u_{j+1} - u_j| &\leq C \fint_{B_{j+1}} |u(z) - u_{j+1}|\, dz \leq C V_{j+1}^{-1} \int_{B_{j+1}} |u(z) - u_{j+1}| dm(z) \\ &\leq C 2^{j(1-\frac{d+1}{2})} \left(\int_{B_{j+1}} |\nabla u(z)|^2 dm(z) \right)^{\frac{1}{2}} \leq C 2^{j\frac{1-d}{2}} \|u\|_W. \end{aligned} \tag{5.41}$$

Since $d > 1$, $(u_j)_{j\in\mathbb{N}}$ is a Cauchy sequence and converges to some value

$$u^0 = \lim_{j\to+\infty} u_j. \tag{5.42}$$

Moreover (5.41) also entails

$$|u_j - u^0| \leq C 2^{j\frac{1-d}{2}} \|u\|_W. \tag{5.43}$$

Let us prove additional properties on u^0. Set $v = |u|$. Notice that

$$|u_j| \leq v_j := \fint_{B_j} |u(z)|\, dz \leq |u_j| + \fint_{B_j} |u(z) - u_j|\, dz \leq |u_j| + C 2^{j\frac{1-d}{2}} \|u\|_W, \tag{5.44}$$

where the last inequality follows from (5.41) (with $j-1$). As a consequence, for any $j \geq 1$, $|v_j - |u_j|| \leq C 2^{j\frac{1-d}{2}} \|u\|_W$ and by taking the limit as $j \to +\infty$,

$$|u^0| = \lim_{j\to+\infty} \fint_{B_j} |u(z)|\, dz. \tag{5.45}$$

In addition,

$$\begin{aligned} &\left| \fint_{B_j} |u(z)|\, dz - |u^0| \right| \\ &\quad \leq |v_j - |u_j|| + ||u_j| - |u^0|| \leq |v_j - |u_j|| + |u_j - u^0| \leq C 2^{j\frac{1-d}{2}} \|u\|_W. \end{aligned} \tag{5.46}$$

Let us show that $\|.\|_W$ is a norm for W^0. Let $u \in W^0$ be such that $\|u\|_W = 0$, then since $W^0 \subset W$, $u \equiv c$ is a constant function. Yet, observe that in this case, $u^0 = c$. The assumption $u \in W^0$ forces $u \equiv c \equiv 0$, that is $\|.\|_W$ is a norm on W^0.

We now prove that $(W^0, \|.\|_W)$ is complete. Let $(v_k)_{k\in\mathbb{N}}$ be a Cauchy sequence in W^0. Since $(v_k - v_l)^0 = 0$, we deduce from (5.46) that for $j \geq 1$ and $k, l \in \mathbb{N}$,

$$\fint_{B_j} |v_k(z) - v_l(z)|\, dz \leq C2^{j\frac{1-d}{2}} \|v_k - v_l\|_W. \tag{5.47}$$

Consequently, $(v_k)_{k\in\mathbb{N}}$ is a Cauchy sequence in L^1_{loc} and thus there exists $u \in L^1_{loc}(\mathbb{R}^n)$ such that $v_k \to u$ in $L^1_{loc}(\mathbb{R}^n)$. Since $(\nabla v_k)_{k\in\mathbb{N}}$ is also a Cauchy sequence in $L^2(\Omega, w)$, there exists $V \in L^2(\Omega, w)$ such that $\nabla v_k \to V$ in $\in L^2(\Omega, w)$. It follows that v_k and ∇v_k converge in the sense of distribution to respectively u and V, thus u has a distributional derivative in Ω and ∇u equals $V \in L^2(\Omega, w)$. In particular $u \in W$. It remains to check that $u^0 = 0$. Yet, notice that

$$|u^0| \leq \left| u^0 - \fint_{B_j} u(z)\, dz \right| + \left| \fint_{B_j} (u - v_k)(z)\, dz \right| + \left| \fint_{B_j} v_k(z)\, dz \right|. \tag{5.48}$$

The first term and the third term in the right-hand side are bounded by $C2^{j\frac{1-d}{2}}\|u\|_W$ and $C2^{j\frac{1-d}{2}}\|u_k\|_W$ respectively (thanks to (5.43)), the second by $C2^{j\frac{1-d}{2}}\|u - u_k\|_W$ (because of (5.47)). By taking k and j big enough, we can make the right-hand side of (5.48) as small as we want. It follows that $u^0 = |u^0| = 0$ and $u \in W^0$. The completeness of W^0 follows.

It remains to check that the completion of $C_0^\infty(\mathbb{R}^n)$ is W^0. However, it is easy to see that any function u in $C_0^\infty(\mathbb{R}^n)$ satisfies $u^0 = 0$ and thus lies in W^0. Together with the fact that W^0 is complete, we deduce that the completion of $C_0^\infty(\mathbb{R}^n)$ with the norm $\|.\|_W$ is included in W^0. The converse inclusion will hold once we establish that any function in W^0 can be approached in the W-norm by functions in $C_0^\infty(\mathbb{R}^n)$. Besides, thanks to Lemma 5.3, it is enough to prove that $u \in W^0$ can be approximated by functions in W that are compactly supported in $\mathbb{R}^n$.

Fix $\phi \in C^\infty((-\infty, +\infty))$ such that $\phi \equiv 1$ on $(-\infty, 1/2]$, $\phi \equiv 0$ on $[1, +\infty)$. For $R > 0$ define ϕ_R by $\phi_R(x) = \phi(\ln|x|/\ln R)$. Observe that that $\phi_R(x) \equiv 1$ if $|x| \leq \sqrt{R}$, $\phi_R(x) \equiv 0$ if $|x| \geq R$ and, for any $x \in \mathbb{R}^n$,

$$|\nabla \phi_R(x)| \leq \frac{C}{\ln R} \frac{1}{|x|}. \tag{5.49}$$

The approximating functions will be the $\phi_R u$, which are compactly supported in $\mathbb{R}^n$. Now

$$\begin{aligned} \|u\phi_R - u\|_W^2 &= \|u(1 - \phi_R)\|_W^2 \\ &\leq \int_\Omega (1 - \phi_R(z))^2 |\nabla u(z)|^2 dm(z) + \int_\Omega |u(z)|^2 |\nabla \phi_R(z)|^2 dm(z) \\ &\leq \int_{|z|\geq\sqrt{R}} |\nabla u(z)|^2 dm(z) + \int_\Omega |u(z)|^2 |\nabla \phi_R(z)|^2 dm(z). \end{aligned} \tag{5.50}$$

By the dominated convergence theorem, the first term of the right-hand side above converges to 0 as R goes to $+\infty$. It remains to check that the second term also tends to 0. Set $C_j = B_j \setminus B_{j-1}$. We have if $R > 1$,

(5.51)

$$\begin{aligned}\int_\Omega |u(z)|^2|\nabla\phi_R(z)|^2 dm(z) &\le \frac{C}{|\ln R|^2}\int_{\sqrt{R}<|z|<R}\frac{|u(z)|^2}{|z|^2}dm(z)\\ &\le \frac{C}{|\ln R|^2}\sum_{j=0}^{\log_2 R+1} 2^{-2j}\int_{C_j}|u(z)|^2dm(z)\\ &\le \frac{C}{|\ln R|^2}\sum_{j=0}^{\log_2 R+1} 2^{-2j}\int_{B_j}|u(z)|^2dm(z)\\ &\le \frac{C}{|\ln R|^2}\sum_{j=0}^{\log_2 R+1} 2^{-2j}\left(\int_{B_j}|u(z)-u_j|^2dm(z)+V_j|u_j|^2\right).\end{aligned}$$

Lemma 4.2 gives that $\int_{B_j}|u(z)-u_j|^2dm(z)$ is bounded, up to a harmless constant, by $2^{2j}\int_{B_j}|\nabla u(z)|^2dm(z) \le 2^{2j}\|u\|_W^2$. In addition, $V_j = m(B_j)$ is bounded by $C2^{j(1+d)}$ because of (2.2) and we get that $|u_j|^2 \le 2^{j(1-d)}\|u\|_W$, by (5.43). Hence

(5.52)

$$\begin{aligned}\int_\Omega |u(z)|^2|\nabla\phi_R(z)|^2 dm(z) &\le \frac{C}{|\ln R|^2}\|u\|_W^2\sum_{j=0}^{\log_2 R+1} 2^{-2j}\left(2^{2j}+2^{j(d+1)}2^{j(1-d)}\right)\\ &\le \frac{C}{|\ln R|^2}\|u\|_W^2\sum_{j=0}^{\log_2 R+1} 1 \le \frac{C}{|\ln R|}\|u\|_W^2,\end{aligned}$$

which converges to 0 as R goes to $+\infty$. This concludes the proof of Lemma 5.7. □

As we shall see now, the situation in low dimensions is different, essentially because when $d \le 1$, the constant function 1 can be approximated by functions of $C_0^\infty(\mathbb{R}^n)$.

LEMMA 5.11. *Let $d \le 1$. For any function u in W, we can find a sequence of functions $(u_k)_{k\in\mathbb{N}}$ in $C_0^\infty(\mathbb{R}^n)$ such that u_k converges, in $L^1_{loc}(\mathbb{R}^n)$ and and for the semi-norm $\|.\|_W$, to u.*

REMARK 5.12. The fact that the function 1 can be approached with the norm $\|.\|_W$ by functions in C_0^∞ means that the completion of C_0^∞ with the norm $\|.\|_W$ is not a space of distributions.

We can legitimately say that the completion of C_0^∞ is embedded into the space of distributions $D' = (C_0^\infty)' \supset L^1_{loc}$ if the convergence $u_k \in C_0^\infty \subset L^1_{loc}$ to $u \in W \subset L^1_{loc}$ in the norm $\|.\|_W$ implies, for $\varphi \in C_0^\infty$, that $\int u_k\varphi$ tends to $\int u\varphi$. Take $u_k \in C_0^\infty(\mathbb{R}^n)$ such that u_k tends to 1 in $L^1_{loc}(\mathbb{R}^n)$ and W. Then since $\|.\|_W$ doesn't see the constants, u_k tends to 0 in W; but the convergence of u_k to 1 in $L^1_{loc}(\mathbb{R}^n)$ implies that $\int u_k\varphi$ tends to $\int \varphi \neq 0$ for some function $\varphi \in C_0^\infty(\mathbb{R}^n)$.

PROOF. As before, we may assume that $0 \in \Gamma$. Let us first prove that for $d \le 1$, the constant function 1 (and thus any constant function) is the limit in W and $L^1_{loc}(\mathbb{R}^n)$ of test functions.

Choose $\phi \in C^\infty([0,+\infty))$ such that $\phi \equiv 1$ on $[0,1/2]$ and $\phi \equiv 0$ on $[1,+\infty)$. For $R > 1$, define ψ_R as $\psi_R(x) = \phi(\ln\ln|x|/\ln\ln R)$ if $|x| > 1$ and $\psi_R(x) = 1$ if

$|x| \leq 1$. This cut-off function is famous for being used by Sobolev, and is useful to handle the critical case (that is, for us, $d = 1$). It can be avoided if $d < 1$ but we didn't want to separate the cases $d < 1$ and $d = 1$. Let us return to the proof of the lemma. We have: $\psi_R(x) \equiv 1$ if $|x| \leq \exp(\sqrt{\ln R})$, $\psi_R(x) \equiv 0$ if $|x| \geq R$ and for any $x \in \mathbb{R}^n$ satisfying $|x| > 1$,

$$|\nabla \psi_R(x)| \leq \frac{C}{\ln \ln R} \frac{1}{|x| \ln |x|}. \tag{5.53}$$

It is easy to see that ψ_R converges to 1 in $L^1_{loc}(\mathbb{R}^n)$ as R goes to $+\infty$. We claim that

$$\|\psi_R\|_W \text{ converges to } 0 \text{ as } R \text{ goes to } +\infty. \tag{5.54}$$

Let us prove (5.54). As in Lemma 5.7, we write B_j for $B(0, 2^j)$ and C_j for $B_j \setminus B_{j-1}$. Then for R large,

$$\begin{aligned}
\|\psi_R\|_W^2 &\leq \frac{C}{|\ln \ln R|^2} \int_{2<|z|\leq R} \frac{1}{|z|^2 |\ln |z||^2} dm(z) \\
&\leq \frac{C}{|\ln \ln R|^2} \sum_{j=1}^{+\infty} 2^{-2j} |\ln 2^j|^{-2} \int_{C_j} dm(z) \\
&\leq \frac{C}{|\ln \ln R|^2} \sum_{j=1}^{+\infty} \frac{1}{j^2} 2^{-2j} 2^{j(d+1)} \leq \frac{C}{|\ln \ln R|^2}.
\end{aligned} \tag{5.55}$$

Our claim follows, and it implies that $\|1 - \psi_R\|_W$ tends to 0.

We will prove now that any function in W can be approached by functions in $C_0^\infty(\mathbb{R}^n)$. Let $u \in W$ be given. Let $u^0 = \fint_{B_0} u$ denote the average of u on the unit ball. We have just seen how to approximate u_0 by test functions, so it will be enough to show that $u - u^0$ can be approached by test functions.

For this we shall proceed as in Lemma 5.7. We shall use the product $\psi_R(u-u^0)$, where ψ_R is the same cut-off function as above, and prove that $\psi_R(u - u^0)$ lies in W and

$$\lim_{R \to +\infty} \|(u - u_0)\psi_R\|_W = 0. \tag{5.56}$$

Notice that $\psi_R(u - u^0)$ is compactly supported, and converges (pointwise and in L^1_{loc}) to $u - u_0$. Thus, as soon as we prove (5.56), Lemma 5.3 will allow us to approximate $(u-u_0)\psi_R$ by smooth, compactly supported functions, and the desired approximation result will follow.

As for the proof of (5.56), of course we shall use Poincaré's inequality, and the the key point will be to get proper bounds on differences of averages of u. These will not be as good as before, because now $d \leq 1$, and instead of working directly on the balls B_j we shall use strings of balls D_j that do not contain the origin, so that their overlap is smaller.

Fix any unit vector $\xi \in \partial B(0, 1)$, and consider the balls

$$D = D^\xi = B(\xi, 9/10) \text{ and, for } j \in \mathbb{N},\ D_j = D_j^\xi = B(2^j\xi, \frac{9}{10}2^j). \tag{5.57}$$

We will later use the D_j^ξ to cover our usual annuli C_j, but in the mean time we fix ξ and want estimates on the numbers $m_j = \fint_{D_j} u_j$.

The Poincaré inequality (4.12), applied with with $p = 1$, yields

$$m(D_j)^{-1} \int_{D_j} |u - m_j| dm \leq C2^j \left(m(D_j)^{-1} \int_{D_j} |\nabla u(z)|^2 dm \right)^{\frac{1}{2}}. \tag{5.58}$$

Of course we have a similar estimate on D_{j+1}; observe also that $D_j \cap D_{j+1}$ contains a ball D'_j of radius 2^{j-2} (we may even take it centered at $2^j\xi$); then

$$\begin{aligned} |m_j - m_{j+1}| &= m(D'_j)^{-1} \int_{D'_j} |m_j - m_{j+1}| dm \\ &\leq m(D'_j)^{-1} \int_{D'_j} (|u - m_j| + |u - m_{j+1}|) dm \\ &\leq Cm(D_j)^{-1} \int_{D_j} |u - m_j| dm + Cm(D_{j+1})^{-1} \int_{D_{j+1}} |u - m_j| dm \\ &\leq C2^j \left(m(D_j)^{-1} \int_{D_j \cup D_{j+1}} |\nabla u(z)|^2 dm \right)^{\frac{1}{2}} \end{aligned} \tag{5.59}$$

because $m(D_j) \leq Cm(D'_j)$ (and similarly for $m(D_{j+1})$), since the measure m is doubling by (2.8). By (2.2), $m(D_j) \geq C^{-1} 2^{j(d+1)}$, so (5.59) yields

$$|m_j - m_{j+1}| \leq C2^{-j(d-1)/2} \left(\int_{D_j \cup D_{j+1}} |\nabla u(z)|^2 dm(z) \right)^{\frac{1}{2}} \tag{5.60}$$

The same estimate, run with $B_0 = B(0,1)$ and D_0 whose intersection also contains a large ball, yields

$$|u^0 - m_0| \leq C \left(\int_{B_0 \cup D_0} |\nabla u(z)|^2 dm(z) \right)^{\frac{1}{2}} \leq C \|u\|_W. \tag{5.61}$$

With ξ fixed, the various $D_j \cup D_{j+1}$ have bounded overlap; thus by (5.60) and Cauchy-Schwarz,

$$\begin{aligned} |m_{j+1} - m_0|^2 &\leq C \left(\sum_{i=0}^{j} 2^{-i(d-1)/2} \|\nabla u\|_{L^2(D_j \cup D_{j+1}, w)} \right)^2 \\ &\leq C(j+1) \sum_{i=0}^{j} 2^{i(1-d)} \|\nabla u\|^2_{L^2(D_j \cup D_{j+1}, w)} \leq C(j+1) 2^{j(1-d)} \|u\|^2_W. \end{aligned} \tag{5.62}$$

Here we used our assumption that $d \leq 1$, and we are happy about our trick with the bounded overlap because a more brutal estimate would lead to a factor $(j+1)^2$ that would hurt us soon. Anyway, we add (5.61) and get that for $j \geq 0$,

$$\left| m_j - u^0 \right|^2 \leq C(j+1) 2^{j(1-d)} \|u\|^2_W. \tag{5.63}$$

We are now ready to prove (5.56). Since the first part of the proof gives that $\|u^0 \psi_R\|_W$ tends to 0, we shall assume that $u^0 = 0$ to simplify the estimates. By Lemma 5.4, $(u - u_0)\psi_R = u\psi_R$ lies in W and its gradient is $u\nabla\psi_R + \psi_R \nabla u$. So we just need to show that when R tends to $+\infty$,

$$\|u\psi_R - u\|_W \leq \|(1 - \psi_R)\nabla u\|_{L^2(\Omega, w)} + \|u \nabla \psi_R\|_{L^2(\Omega, w)} \tag{5.64}$$

tends to 0. The first term of the right-hand side converges to 0 as R goes to $+\infty$, thanks to the dominated convergence theorem, and for the second term we use (5.53) and the fact that $\nabla\psi_R$ is supported in the region Z_R where $\exp(\sqrt{\ln R}) \leq |x| \leq R$. Thus

(5.65)
$$\|u\nabla\psi_R\|^2_{L^2(\Omega,w)} = \int_{\mathbb{R}^n} |u(z)|^2|\nabla\psi_R(z)|^2 dm(z) \leq \frac{C}{|\ln\ln R|^2}\int_{Z_R} \frac{|u(z)|^2}{|z|^2(\ln|z|)^2} dm(z)$$

As usual, we cut Z_R into annular subregions C_j, and then further into balls like the D_j. We start with the $C_j = B_j \setminus B_{j-1}$. For R large, if C_j meets Z_R, then $10 \leq j \leq 1 + \log_2 R$ and

$$\int_{C_j} \frac{|u(z)|^2}{|z|^2(\ln|z|)^2} dm(z) \leq j^{-2}2^{-2j}\int_{C_j} |u(z)|^2 dm(z). \tag{5.66}$$

We further cut C_j into balls, because we want to apply Poincaré's inequality. Let the D_j^ξ be as in the definition (5.57). We can find a finite set $\Xi \subset \partial B(0,1)$ such that the balls D^ξ, $\xi \in \Xi$, cover $B(0,1) \setminus B(0,1/2)$. Then for $j \geq 1$ the D^ξ, $\xi \in \Xi$, cover C_j and, by (5.65) and (5.66),

$$\|u\nabla\psi_R\|^2_{L^2(\Omega,w)} \leq \frac{C}{|\ln\ln R|^2}\sum_{j=10}^{1+\log_2 R} j^{-2}2^{-2j}\sum_{\xi\in\Xi}\int_{D_j^\xi} |u(z)|^2 dm(z). \tag{5.67}$$

Then by the Poincaré inequality (4.12) (with $p = 2$),

$$\int_{D_j^\xi} |u(z) - m_j^\xi|^2 dm(z) \leq C2^{2j}\int_{D_j^\xi} |\nabla u(z)|^2 dm(z), \tag{5.68}$$

where $m_j^\xi = \fint_{D_j^\xi}$ as in the estimates above. Thus

$$\int_{D_j^\xi} |u(z)|^2 dm(z) \leq C2^{2j}\int_{D_j^\xi} |\nabla u(z)|^2 dm(z) + Cm(D_j^\xi)(j+1)2^{j(1-d)}\|u\|^2_W \tag{5.69}$$

by (5.63) and because $u^0 = 0$. But $m(D_j^\xi) \leq C2^{(d+1)j}$ by (2.2), so

$$\int_{D_j^\xi} |u(z)|^2 dm(z) \leq C(j+1)2^{2j}\|u\|^2_W. \tag{5.70}$$

We return to (5.67) and get that

$$\begin{aligned}\|u\nabla\psi_R\|^2_{L^2(\Omega,w)} &\leq \frac{C}{|\ln\ln R|^2}\sum_{j=10}^{1+\log_2 R} j^{-2}2^{-2j}\sum_{\xi\in\Xi}(j+1)2^{2j}\|u\|^2_W \\ &\leq \frac{C}{|\ln\ln R|^2}\sum_{j=10}^{1+\log_2 R} j^{-1}\|u\|^2_W \leq \frac{C}{|\ln\ln R|}\|u\|^2_W\end{aligned} \tag{5.71}$$

because Ξ is finite, and where we see that j^{-1} is really useful.

We already took care of the other part of (5.64); thus $\|u\psi_R - u\|_W$ tends to 0. This proves (5.56) (recall that $u^0 = 0$), and completes our proof of Lemma 5.11. □

CHAPTER 6

The Chain Rule and Applications

We record here some basic (and not shocking) properties concerning the derivative of $f \circ u$ when $u \in W$, and the fact that $uv \in W \cap L^\infty$ when $u, v \in W \cap L^\infty$.

LEMMA 6.1. *The following properties hold:*

(a) *Let $f \in C^1(\mathbb{R})$ be such that f' is bounded and let $u \in W$. Then $f \circ u \in W$ and*

$$\nabla(f \circ u) = f'(u)\nabla u. \tag{6.1}$$

Moreover, $T(f \circ u) = f \circ (Tu)$ σ-a.e. in Γ.

(b) *Let $u, v \in W$. Then $\max\{u, v\}$ and $\min\{u, v\}$ belong to W and, for almost every $x \in \mathbb{R}^n$,*

$$\nabla \max\{u,v\}(x) = \begin{cases} \nabla u(x) & \text{if } u(x) \geq v(x) \\ \nabla v(x) & \text{if } v(x) \geq u(x) \end{cases} \tag{6.2}$$

and

$$\nabla \min\{u,v\}(x) = \begin{cases} \nabla u(x) & \text{if } u(x) \leq v(x) \\ \nabla v(x) & \text{if } v(x) \leq u(x). \end{cases} \tag{6.3}$$

In particular, for any $\lambda \in \mathbb{R}$, $\nabla u = 0$ (Lebesgue) a.e. on $\{x \in \mathbb{R}^n,\ u(x) = \lambda\}$.

In addition, $T \max\{u,v\} = \max\{Tu, Tv\}$ and $T \min\{u,v\} = \min\{Tu, Tv\}$ σ-a.e. on Γ. Thus $\max\{u,v\}$ and $\min\{u,v\}$ lie in W_0 as soon as $u, v \in W_0$.

REMARK 6.2. A consequence of Lemma 6.1 (b) is that, for example, $|u| \in W$ (resp. $|u| \in W_0$) whenever $u \in W$ (resp. $u \in W_0$).

PROOF. A big part of this proof follows the results from 1.18 to 1.23 in [**HKM**].

Let us start with (a). More precisely, we aim for (6.1). Let $f \in C^1(\mathbb{R}) \cap Lip(\mathbb{R})$ and let $u \in W$. The idea of the proof is the following: we approximate u by smooth functions φ_k, for which the result is immediate. Then we observe that both $\nabla(f \circ u)$ and $f'(u)\nabla u$ are the limit (in the sense of distributions) of the gradient of $f \circ \varphi_k$.

According to Lemma 5.3, there exists a sequence $(\varphi_k)_{k \in \mathbb{N}}$ of functions in $C^\infty(\mathbb{R}^n) \cap W$ such that $\varphi_k \to u$ in W and in $L^1_{loc}(\mathbb{R}^n)$. The classical (thus distributional) derivative of $f \circ \varphi_k$ is

$$\nabla[f \circ \varphi_k] = f'(\varphi_k)\nabla\varphi_k. \tag{6.4}$$

In particular, since $\varphi_k \in W$ and f' is bounded, $f \circ \varphi_k \in W$ and $\|f \circ \varphi_k\|_W \leq \|\varphi_k\|_W \sup |f'|$.

Notice that $|f(s)-f(t)| \leq |s-t| \sup|f'|$. Therefore, since $\varphi_k \to u$ in $L^1_{loc}(\mathbb{R}^n)$, for any ball $B \subset \mathbb{R}^n$

$$\int_B |f\circ\varphi_k - f\circ u|\,dx \leq \sup|f'| \int_B |\varphi_k - u|\,dx \longrightarrow 0. \tag{6.5}$$

That is $f\circ\varphi_k \to f\circ u$ in $L^1_{loc}(\mathbb{R}^n)$, hence also in the sense of distributions. Besides,

$$\begin{aligned}
&\left(\int_\Omega |f'(\varphi_k)\nabla\varphi_k - f'(u)\nabla u|^2 dm\right)^{\frac{1}{2}} \\
&\quad \leq \left(\int_\Omega |f'(\varphi_k)[\nabla\varphi_k - \nabla u]|^2 dm\right)^{\frac{1}{2}} + \left(\int_\Omega |\nabla u[f'(\varphi_k) - f'(u)]|^2 dm\right)^{\frac{1}{2}} \\
&\quad \leq \sup|f'| \left(\int_\Omega |\nabla\varphi_k - \nabla u|^2 dm\right)^{\frac{1}{2}} + \left(\int_\Omega |\nabla u|^2 |f'(\varphi_k) - f'(u)|^2 dm\right)^{\frac{1}{2}}.
\end{aligned} \tag{6.6}$$

The first term in the right-hand side is converges to 0 since $\varphi_k \to u$ in W. Besides, $\varphi_k \to u$ a.e. in Ω and f' is continuous, so $f'(\varphi_k) \to f'(u)$ a.e. in Ω. Therefore, the second term also converges to 0 thanks to the dominated convergence theorem. It follows that $\nabla[f\circ\varphi_k] \to f'(u)\nabla u$ in $L^2(\Omega, w)$, and hence also in the sense of distributions. We proved that $f\circ\varphi_k \to f\circ u$ and $\nabla[f\circ\varphi_k] \to f'(u)\nabla u \in L^2(\Omega, w)$ in the sense of distributions, and so the distributional derivative of $f\circ u$ lies in $L^2(\Omega, w)$ and is equal to $f'(u)\nabla u$. In particular, $f\circ u \in W$. Note that we also proved that $f\circ\varphi_k \to f\circ u$ in W.

In order to finish the proof of (a), we need to prove that $T(f\circ u) = f(Tu)$ σ-a.e. in Γ. If $v \in W$ is also a continuous function on $\mathbb{R}^n$, then it is easy to check from the definition of the trace that $Tv(x) = v(x)$ for every $x \in \Gamma$. Since $f\circ\varphi_k$ and φ_k are both continuous functions, we get that

$$f\circ\varphi_k(x) = T(f\circ\varphi_k)(x) = f(T\varphi_k(x)) \quad \text{for } x \in \Gamma \text{ and } k \in \mathbb{N}. \tag{6.7}$$

Hence for every ball B centered on Γ and every $k \geq 0$,

$$\begin{aligned}
\int_B |T(f\circ u) - f(Tu)|d\sigma &\leq \int_B |T(f\circ u) - T(f\circ\varphi_k)|d\sigma + \int_B |f(T\varphi_k) - f(Tu)|d\sigma \\
&\leq \int_B |T(f\circ u) - T(f\circ\varphi_k)|d\sigma + \sup|f'| \int_B |T\varphi_k - Tu|d\sigma.
\end{aligned} \tag{6.8}$$

Recall that each convergence $\varphi_k \to u$ and $f\circ\varphi_k \to f\circ u$ holds in both W and $L^1_{loc}(\mathbb{R}^n)$. The assertion (5.16) then gives that both convergences $T\varphi_k \to Tu$ and $T(f\circ\varphi_k) \to T(f\circ u)$ hold in $L^1_{loc}(\Gamma, \sigma)$. Thus the right-hand side of (6.8) converges to 0 as k goes to $+\infty$. We obtain that for every ball B centered on Γ,

$$\int_B |T(f\circ u) - f(Tu)|d\sigma = 0; \tag{6.9}$$

in particular, $T(f\circ u) = f(Tu)$ σ-a.e. in Γ.

Let us turn to the proof of (b). Set $u^+ = \max\{u, 0\}$. Then $\max\{u, v\} = (u-v)^+ + v$ and $\min\{u, v\} = u - (u-v)^+$. Thus is it enough to show that for any $u \in W$, u^+ lies in W and satisfies

$$\nabla u^+(x) = \begin{cases} \nabla u(x) & \text{if } u(x) > 0 \\ 0 & \text{if } u(x) \leq 0 \end{cases} \quad \text{for almost every } x \in \mathbb{R}^n \tag{6.10}$$

and

$$T(u^+) = (Tu)^+ \quad \sigma\text{-almost everywhere on } \Gamma. \tag{6.11}$$

Note that in particular (6.10) implies that $\nabla u = 0$ a.e. in $\{u = \lambda\}$. Indeed, since $u = \lambda + (u-\lambda)_+ - (\lambda - u)_+$, (6.10) implies that for almost every $x \in \Omega$,

$$\nabla u(x) = \begin{cases} \nabla u(x) & \text{if } u(x) \neq \lambda \\ 0 & \text{if } u(x) = \lambda. \end{cases} \tag{6.12}$$

Let us prove the claim (6.10). Define f and $g = \mathbb{1}_{(0,+\infty)}$ by $f(t) = \max\{0,t\}$ and $g(t) = 0$ when $t \leq 0$ and $g(t) = 1$ when $t > 0$. Our aim is to approximate f by an increasing sequence of C^1-functions and then to conclude by using (a) and the monotone convergence theorem. Define for any integer $j \geq 1$ the function f_j by

$$f_j(t) = \begin{cases} 0 & \text{if } t \leq 0 \\ \frac{j}{j+1} t^{\frac{j+1}{j}} & \text{if } 0 \leq t \leq 1 \\ t - \frac{1}{j+1} & \text{if } t \geq 1. \end{cases} \tag{6.13}$$

Notice that f_j is non-negative and (f_j) is a nondecreasing sequence that converges pointwise to f. Consequently, $f_j \circ u \geq 0$ and $(f_j \circ u)$ is a nondecreasing sequence that converges pointwise to $f \circ u = u^+ \in L^1_{loc}(\mathbb{R}^n)$. The monotone convergence theorem implies that $f_j \circ u \to u^+$ in $L^1_{loc}(\mathbb{R}^n)$.

Moreover, f_j lies in $C^1(\mathbb{R})$ and its derivative is

$$f_j'(t) = \begin{cases} 0 & \text{if } t \leq 0 \\ t^{\frac{1}{j}} & \text{if } 0 \leq t \leq 1 \\ 1 & \text{if } t \geq 1. \end{cases} \tag{6.14}$$

Thus f_j' is bounded and part (a) of the lemma implies $f_j \circ u \in W$ and $\nabla(f_j \circ u) = f_j'(u)\nabla u$ almost everywhere on $\mathbb{R}^n$. In addition, f_j' converges to g pointwise everywhere, so

$$\nabla(f_j \circ u) = f_j'(u)\nabla u \to v := (g \circ u)\nabla u = \begin{cases} \nabla u & \text{if } u > 0 \\ 0 & \text{if } u \leq 0 \end{cases} \tag{6.15}$$

almost everywhere (i.e., wherever $\nabla(f_j \circ u) = f_j'(u)\nabla u$). The convergence also occurs in $L^2(\Omega, w)$ and in $L^1_{loc}(\mathbb{R}^n)$, because $|\nabla(f_j \circ u)| \leq |\nabla u|$ and by the dominated convergence theorem, and therefore also in the sense of distribution. Since $f_j \circ u$ converges to u^+ pointwise almost everywhere and hence (by the dominated convergence theorem again) in L^1_{loc} and in the sense of distributions, we get that $v = (g \circ u)\nabla u$ is the distribution derivative of u^+. This completes the proof of (6.10).

Finally, let us establish (6.11). The plan is to prove that we can find smooth functions φ_k such that φ_k^+ converges (in $L^1_{loc}(\Gamma, \sigma)$) to both Tu^+ and $(Tu)^+$. We claim that for $u \in W$ and any sequence (u_k) in W, the following implication holds true:

$$u_k \to u \text{ pointwise a.e. and in } W \implies u_k^+ \to u^+ \text{ pointwise a.e. and in } W. \tag{6.16}$$

First we assume the claim and prove (6.11). With the help of Lemma 5.3, take $(\varphi_k)_{k\in\mathbb{N}}$ be a sequence of functions in $C^\infty(\mathbb{R}^n)$ such that $\varphi_k \to u$ in W, and in $L^1_{loc}(\mathbb{R}^n)$. We may also replace (φ_k) by a subsequence, and get that $\varphi_k \to u$ pointwise a.e. The claim (6.16) implies that $\varphi_k^+ \to u^+$ in W. In addition, $\varphi_k^+ \to u^+$

in $L^1_{loc}(\mathbb{R}^n)$, for instance because φ_k tends to u in $L^1_{loc}(\mathbb{R}^n)$ and by the estimate $|\varphi_k^+ - u^+| \leq |\varphi_k - u|$.

Thus we may apply (5.16), and we get that $T\varphi_k^+$ tends to Tu^+ in $L^1_{loc}(\Gamma)$. Since φ_k^+ is continuous, $T\varphi_k^+ = \varphi_k^+$ and

$$\varphi_k^+ \text{ tends to } Tu^+ \text{ in } L^1_{loc}(\Gamma). \tag{6.17}$$

We also need to check that φ_k^+ converges to $(Tu)^+$. Notice that (5.16) also implies that $\varphi_k \to Tu$ in $L^1_{loc}(\Gamma, \sigma)$. Together with the easy fact that $|a^+ - b^+| \leq |a - b|$ for $a, b \in \mathbb{R}$, this proves that $\varphi_k^+ \to (Tu)^+$ in $L^1_{loc}(\Gamma, \sigma)$.

We just proved that φ_k^+ converges to both $T(u^+)$ and $(Tu)^+$ in $L^1_{loc}(\Gamma, \sigma)$. By uniqueness of the limit, $T(u^+) = (Tu)^+$ σ-a.e. in Γ, as needed for (6.11). Thus the proof of the lemma will be complete as soon as we establish the claim (6.16).

First notice that $|u_j^+ - u^+| \leq |u_j - u|$ and thus the a.e. pointwise convergence of $u_j \to u$ yields the a.e. pointwise convergence $u_j^+ \to u^+$. Let g denote the characteristic function of $(0, +\infty)$; then by (6.10)

(6.18)
$$\begin{aligned}\left(\int_\Omega |\nabla u_j^+ - \nabla u^+|^2 dm\right)^{\frac{1}{2}} &= \left(\int_\Omega |g(u_j)\nabla u_j - g(u)\nabla u|^2 dm\right)^{\frac{1}{2}} \\ &\leq \left(\int_\Omega |g(u_j)[\nabla u_j - \nabla u]|^2 dm\right)^{\frac{1}{2}} + \left(\int_\Omega |\nabla u[g(u_j) - g(u)]|^2 dm\right)^{\frac{1}{2}} \\ &\leq \left(\int_\Omega |\nabla u_j - \nabla u|^2 dm\right)^{\frac{1}{2}} + \left(\int_\Omega |\nabla u|^2 |g(u_j) - g(u)|^2 dm\right)^{\frac{1}{2}}.\end{aligned}$$

The first term in the right-hand side converges to 0 since $u_j \to u$ in W. Call I the second term; I is finite, since $u \in W$ and $|g(u_j) - g(u)| \leq 1$. Moreover, thanks to (6.12), $\nabla u = 0$ a.e. on $\{u = 0\}$. So the square of I can be written

$$I^2 = \int_{\{|u|>0\}} |\nabla u|^2 |g(u_j) - g(u)|^2 dm. \tag{6.19}$$

Let $x \in \{|u| > 0\}$ be such that $u_j(x)$ converges to $u(x) \neq 0$; then there exists $j_0 \geq 0$ such that for $j \geq j_0$ the sign of $u_j(x)$ is the same as the sign of $u(x)$. That is, $g(u_j)(x)$ converges to $g(u)(x)$. Since $u_j \to u$ a.e. in Ω, the previous argument implies that $g(u_j) \to g(u)$ a.e. in $\{|u| > 0\}$. Then I^2 converges to 0, by the dominated convergence theorem. Going back to (6.18), we obtain that $u_j^+ \to u^+$ in W, which concludes our proof of (6.16); Lemma 6.1 follows. □

LEMMA 6.3. *Let $u, v \in W \cap L^\infty(\Omega)$. Then $uv \in W \cap L^\infty(\Omega)$, $\nabla(uv) = v\nabla u + u\nabla v$, and $T(uv) = Tu \cdot Tv$.*

PROOF. Without loss of generality, we can assume that $\|u\|_\infty, \|v\|_\infty \leq 1$. The fact that $uv \in L^\infty(\Omega)$ is immediate. Let us now prove that $uv \in W$. According to Lemma 5.3, there exists two sequences $(u_j)_{j\in\mathbb{N}}$ and $(v_j)_{j\in\mathbb{N}}$ of functions in $C^\infty(\mathbb{R}^n) \cap W$ such that $u_j \to u$ and $v_j \to v$ in W, in $L^1_{loc}(\mathbb{R}^n)$, and pointwise. Besides, the construction of u_j, v_j given by Lemma 5.3 allows us to assume that $\|u_j\|_\infty \leq \|u\|_\infty \leq 1$ and $\|v_j\|_\infty \leq \|v\|_\infty \leq 1$. The distributional derivative of $u_j v_j$ equals the classical derivative, which is

$$\nabla(u_j v_j) = v_j \nabla u_j + u_j \nabla v_j. \tag{6.20}$$

Since u_j and v_j lie in W, (6.20) says that $u_jv_j \in W$. The bound

(6.21)
$$\int_B |u_jv_j - uv| \leq \int_B |u_j||v_j - v| + \int_B |v||u_j - u| \leq \|v_j - v\|_{L^1(B)} + \|u_j - u\|_{L^1(B)},$$

which holds for any ball $B \subset \mathbb{R}^n$, shows that $u_jv_j \to uv$ in $L^1_{loc}(\mathbb{R}^n)$. Moreover,

(6.22)
$$\begin{aligned}
&\left(\int_B |(u_j\nabla v_j + v_j\nabla u_j) - (u\nabla v + v\nabla u)|^2 dm\right)^{\frac{1}{2}} \\
&\qquad \leq \left(\int_B |u_j\nabla v_j - u\nabla v|^2 dm\right)^{\frac{1}{2}} + \left(\int_B |v_j\nabla u_j - v\nabla u|^2 dm\right)^{\frac{1}{2}} \\
&\qquad \leq \left(\int_B |u_j|^2|\nabla v_j - \nabla v|^2 dm\right)^{\frac{1}{2}} + \left(\int_B |u_j - u|^2|\nabla v|^2 dm\right)^{\frac{1}{2}} \\
&\qquad\qquad + \left(\int_B |v_j|^2|\nabla u_j - \nabla u|^2 dm\right)^{\frac{1}{2}} + \left(\int_B |v_j - v|^2|\nabla u|^2 dm\right)^{\frac{1}{2}}.
\end{aligned}$$

The first and third terms in the right-hand side converge to 0 as j goes to $+\infty$, because $|u_j|, |v_j| \leq 1$ and since $u_j \to u$ and $v_j \to v$ in W. The second and forth terms also converge to 0 thanks to the dominated convergence theorem (and the fact that $u_j \to u$ and $v_j \to v$ pointwise a.e.). We deduce that $\nabla(u_jv_j) = u_j\nabla v_j + v_j\nabla u_j \to u\nabla v + v\nabla u$ in $L^2(\Omega, w)$. By the uniqueness of the distributional derivative, $\nabla(uv) = u\nabla v + v\nabla u \in L^2(\Omega, w)$. In particular, $uv \in W$. Note that we also proved that $u_jv_j \to uv$ in W.

It remains to prove that $T(uv) = Tu \cdot Tv$. Since u_jv_j is continuous and $u_jv_j \to uv$ in W and $L^1_{loc}(\mathbb{R}^n)$, then by (5.16), $u_jv_j = T(u_jv_j) \to T(uv)$ in $L^1_{loc}(\Gamma, \sigma)$. Moreover, for any ball B centered on Γ,

$$\begin{aligned}
\int_B |u_jv_j - Tu \cdot Tv| d\sigma &\leq \int_B |u_j||v_j - Tv| d\sigma + \int_B |u_j - Tu||Tv| d\sigma \\
&\leq \int_B |v_j - Tv| d\sigma + \int_B |u_j - Tu| d\sigma
\end{aligned} \tag{6.23}$$

where the last line holds because $|u_j| \leq 1$ and $|Tv| \leq \sup |v| \leq 1$, where the later bound either follows from Lemma 6.1 or is easily deduced from the definition of the trace. By construction, $u_j \to u$ and $v_j \to v$ in W and $L^1_{loc}(\mathbb{R}^n)$. Then by (5.16) the right-hand side terms in (6.23) converge to 0.

We proved that u_jv_j converges in $L^1_{loc}(\Gamma, \sigma)$ to both $T(uv)$ and $Tu \cdot Tv$. By uniqueness of the limit, $T(uv) = Tu \cdot Tv$ σ-a.e. in Γ. Lemma 6.3 follows. $\square$

CHAPTER 7

The Extension Operator

The main point of this chapter is the construction of our extension operator $E : H \to W$, which will be done naturally, with the Whitney extension scheme that uses dyadic cubes.

Our main object will be a function g on Γ, that typically lies in H or in $L^1_{loc}(\Gamma, \sigma)$. We start with the Lebesgue density result for $g \in L^1_{loc}(\Gamma, \sigma)$ that was announced in the introduction.

LEMMA 7.1. *For any $g \in L^1_{loc}(\Gamma, \sigma)$ and σ-almost all $x \in \Gamma$,*

$$\lim_{r \to 0} \fint_{\Gamma \cap B(x,r)} |g(y) - g(x)| d\sigma(y) = 0. \tag{7.1}$$

PROOF. Since (Γ, σ) satisfies (1.1), the space (Γ, σ) equipped with the metric induced by $\mathbb{R}^n$ is a doubling space. Indeed, let B be a ball centered on Γ. According to (1.1),

$$\sigma(2B) \leq 2^d C_0 r^d \leq 2^d C_0^2 r^d \sigma(B). \tag{7.2}$$

From there, the lemma is only a consequence of the Lebesgue differentiation theorem in doubling spaces (see for example [**Fed**, Sections 2.8-2.9]). □

REMARK 7.2. We claim that $H \subset L^1_{loc}(\Gamma, \sigma)$, and hence (7.1) holds for $g \in H$ and σ-almost every $x \in \Gamma$. Indeed, let B be a ball centered on Γ, then a brutal estimate yields

$$\begin{aligned} \int_B \int_B |g(x) - g(y)| d\sigma(x) d\sigma(y) &\leq C_B \left(\int_B \int_B |g(x) - g(y)|^2 d\sigma(x) d\sigma(y) \right)^{\frac{1}{2}} \\ &\leq C_B \|g\|_H < +\infty. \end{aligned} \tag{7.3}$$

Hence for σ-almost every $x \in B \cap \Gamma$, $\int_B |g(x) - g(y)| d\sigma(y) < +\infty$. In particular, since $\sigma(B) > 0$, there exists $x \in B \cap \Gamma$ such that $\int_B |g(x) - g(y)| d\sigma(y) < +\infty$. We get that $g \in L^1(B, \sigma)$, and our claim follows.

Let us now start the construction of the extension operator $E : H \to W$. We proceed as for the Whitney extension theorem, with only a minor modification because averages will be easier to manipulate than specific values of g.

We shall use the family $\mathcal{W}$ of dyadic Whitney cubes constructed as in the first pages of [**Ste**] and the partition of unity $\{\varphi_Q\}, Q \in \mathcal{W}$, that is usually associated to $\mathcal{W}$. Recall that $\mathcal{W}$ is the family of maximal dyadic cubes Q (for the inclusion) such that $20Q \subset \Omega$, say, and the φ_Q are smooth functions such that φ_Q is supported in $2Q$, $0 \leq \varphi_Q \leq 1$, $|\nabla \varphi_Q| \leq C \mathrm{diam}(Q)^{-1}$, and $\sum_Q \varphi_Q = \mathbb{1}_\Omega$.

Let us record a few of the simple properties of $\mathcal{W}$. These are simple, but yet we refer to [**Ste**, Chapter VI] for details. It will be convenient to denote by $r(Q)$

the side length of the dyadic cube Q. Also set $\delta(Q) = \operatorname{dist}(Q, \Gamma)$. For $Q \in \mathcal{W}$, we select a point $\xi_Q \in \Gamma$ such that $\operatorname{dist}(\xi_Q, Q) \leq 2\delta(Q)$, and set

$$B_Q = B(\xi_Q, \delta(Q)). \tag{7.4}$$

If $Q, R \in \mathcal{W}$ are such that $2Q$ meets $2R$, then $r(R) \in \{\frac{1}{2}r(Q), r(Q), 2r(Q)\}$; then we can easily check that $R \subset 8Q$. Thus R is a dyadic cube in $8Q$ whose side length is bigger than $\frac{1}{2}r(Q)$; there exist at most $2 \cdot 16^n$ dyadic cubes like this. This proves that there is a constant $C = C(n)$ such that for $Q \in \mathcal{W}$,

$$\text{the number of cubes } R \in \mathcal{W} \text{ such that } 2R \cap 2Q \neq \emptyset \text{ is at most } C. \tag{7.5}$$

The operator E is defined on functions in $L^1_{loc}(\Gamma, \sigma)$ by

$$Eg(x) = \sum_{Q \in \mathcal{W}} \varphi_Q(x) y_Q, \tag{7.6}$$

where we set

$$y_Q = \fint_{B_Q} g(z) d\sigma(z), \tag{7.7}$$

with B_Q as in (7.4). For the extension of Lipschitz functions, for instance, one would take $y_Q = g(\xi_Q)$, but here we will use the extra regularity of the averages.

Notice that Eg is continuous on Ω, because the sum in (7.6) is locally finite. Indeed, if $x \in \Omega$ and $Q \in \mathcal{W}$ contains x, (7.5) says that there are at most C cubes $R \in \mathcal{W}$ such that φ_R does not vanish on $2Q$; then the restriction of Eg to $2Q$ is a finite sum of continuous functions. Moreover, if g is continuous on Γ, then Eg is continuous on the whole $\mathbb{R}^n$. We refer to [**Ste**, Proposition VI.2.2] for the easy proof.

THEOREM 7.3. *For any $g \in L^1_{loc}(\Gamma, \sigma)$ (and by Remark 7.2, this applies to $g \in H$),*

$$T(Eg) = g \quad \sigma\text{-a.e. in } \Gamma. \tag{7.8}$$

Moreover, there exists $C > 0$ such that for any $g \in H$,

$$\|Eg\|_W \leq C\|g\|_H. \tag{7.9}$$

PROOF. Let $g \in L^1_{loc}$ be given, and set $u = Eg$. We start the proof with the verification of (7.8). Recall that by definition of the trace,

$$T(E(g))(x) = \lim_{r \to 0} \fint_{B(x,r)} u(z)\, dz \tag{7.10}$$

for σ-almost every $x \in \Gamma$; we want to prove that this limit is $g(x)$ for almost every $x \in \Gamma$, and we can restrict to the case when x is a Lebesgue point for g (as in (7.1)).

Fix such an $x \in \Gamma$ and $r > 0$. Set $B = B(x, r)$, then

$$\left| \fint_{B(x,r)} u(z)\, dz - g(x) \right| \leq \fint_B |u(z) - g(x)| dz \leq Cr^{-n} \sum_{R \in \mathcal{W}(B)} \int_R |u(z) - g(x)| dz, \tag{7.11}$$

where we denote by $\mathcal{W}(B)$ the set of cubes $R \in \mathcal{W}$ that meet B.

Let $R \in \mathcal{W}$ and $z \in R$ be given. Recall from (7.6) that $u(z) = \sum_{Q \in \mathcal{W}} \varphi_Q(z) y_Q$; the sum has less than C terms, corresponding to cubes $Q \in \mathcal{W}$ such that $z \in 2Q$. If

Q is such a cube, we have seen that $\frac{1}{2}r(R) \leq r(Q) \leq 2r(R)$, and since $\delta(R) \geq 10r(Q)$ because $20Q \subset \Omega$, a small computation with (7.4) yields that $B_Q \subset 100B_R$. Hence

$$(7.12)\quad |y_Q - g(x)| = \Big| \fint_{B_Q} g\,d\sigma - g(x)\Big| \leq \fint_{B_Q} |g-g(x)|d\sigma \leq C\fint_{100B_R} |g-g(x)|d\sigma.$$

Since $u(z)$ is an average of such y_Q, we also get that $|u(z) - g(x)| \leq C\fint_{100B_R} |g - g(x)|d\sigma$, and (7.11) yields

$$(7.13)\qquad \Big|\fint_{B(x,r)} u\,dz - g(x)\Big| \leq Cr^{-n} \sum_{R\in\mathcal{W}(B)} |R| \fint_{100B_R} |g - g(x)|d\sigma.$$

Notice that $\delta(R) = \operatorname{dist}(R,\Gamma) \leq \operatorname{dist}(R,x) \leq r$ because R meets $B = B(x,r)$ and $x \in \Gamma$, so, by definition of $\mathcal{W}$, the sidelength of R is such that $r(R) \leq Cr$. Let $\mathcal{W}_k(B)$ be the collection of $R \in \mathcal{W}(B)$ such that $r(R) = 2^k$. For each k, the balls $100B_R$, $R \in \mathcal{W}_k(B)$ have bounded overlap (because the cubes R are essentially disjoint and they have the same sidelength), and they are contained in $B' = B(x, Cr)$. Thus

$$\begin{aligned}\sum_{R\in\mathcal{W}_k(B)} |R| \fint_{100B_R} |g - g(x)|d\sigma &\leq C2^{nk}2^{-dk} \sum_{R\in\mathcal{W}_k(B)} \int_{100B_R} |g - g(x)|d\sigma \\ (7.14)\qquad &\leq C2^{(n-d)k} \int_{B'} |g - g(x)|d\sigma.\end{aligned}$$

We may sum over k (because $2^k = r(R) \leq Cr$ when $R \in \mathcal{W}_k(B)$, and the exponent $n - d$ is positive). We get that

(7.15)

$$\Big|\fint_{B(x,r)} u\,dz - g(x)\Big| \leq Cr^{-n}\sum_k 2^{(n-d)k} \int_{B'} |g - g(x)|d\sigma \leq Cr^{-d} \int_{B'} |g - g(x)|d\sigma.$$

If x is a Lebesgue point for g, (7.1) says that both sides of (7.15) tend to 0 when r tends to 0. Recall from (7.10) that for almost every $x \in \Gamma$, $T(E(g))(x)$ is the limit of $\fint_{B(x,r)} u$; if in addition x is a Lebesgue point, we get that $T(E(g))(x) = g(x)$. This completes our proof of (7.8).

Now we show that for $g \in H$, $u \in W$ and even $\|u\|_W \leq C\|g\|_H$. The fact that u is locally integrable in Ω is obvious (u is continuous there because the cubes $2Q$ have bounded overlap), and similarly the distribution derivative is locally integrable, and given by

$$(7.16)\qquad \nabla u(x) = \sum_{Q\in\mathcal{W}} y_Q \nabla\varphi_Q(x) = \sum_{Q\in\mathcal{W}} [y_Q - y_R]\nabla\varphi_Q(x),$$

where in the second part (which will be used later) we can pick for R any given cube (that may depend on x), for instance, one to the cubes of $\mathcal{W}$ that contains x, and the identity holds because $\sum_Q \nabla\varphi_Q = \nabla(\sum_Q \varphi_Q) = 0$. Thus the question is merely the computation of

$$\begin{aligned}\|u\|_W^2 &= \int_\Omega |\nabla u(x)|^2 w(x)dx = \sum_{R\in\mathcal{W}} \int_R |\nabla u(x)|^2 w(x)dx \\ (7.17)\qquad &\leq C\sum_{R\in\mathcal{W}} \delta(R)^{d+1-n} \int_R |\nabla u(x)|^2 dx\end{aligned}$$

(because $w(x) = \delta(x)^{d+1-n} \leq \delta(R)^{d+1-n}$ when $x \in R$). Fix $R \in \mathcal{W}$, denote by $\mathcal{W}(R)$ the set of cubes $Q \in \mathcal{W}$ such that $2Q$ meets R, and observe that for $x \in R$,

$$|\nabla u(x)| \leq \sum_{Q \in \mathcal{W}(R)} \big|[y_Q - y_R]\nabla\varphi_Q(x)\big| \leq C\delta(R)^{-1} \sum_{Q \in \mathcal{W}(R)} \big|y_Q - y_R\big| \tag{7.18}$$

because $|\nabla\varphi_Q(x)| \leq C\delta(Q)^{-1} \leq C\delta(R)^{-1}$ by definitions and the standard geometry of Whitney cubes. In turn,

$$\begin{aligned} \big|y_Q - y_R\big| &\leq \fint_{\Gamma\cap B_Q}\fint_{\Gamma\cap B_R} |g(x) - g(y)| d\sigma(x) d\sigma(y) \\ &\leq \Big\{\fint_{\Gamma\cap B_Q}\fint_{\Gamma\cap B_R} |g(x) - g(y)|^2 d\sigma(x) d\sigma(y)\Big\}^{1/2} \\ &\leq C\delta(R)^{-d}\Big\{\int_{\Gamma\cap B_R}\int_{\Gamma\cap 100B_R} |g(x) - g(y)|^2 d\sigma(x) d\sigma(y)\Big\}^{1/2} \end{aligned} \tag{7.19}$$

by (1.1) and because $B_Q \subset 100B_R$. Thus by (7.18)

$$\begin{aligned} \int_R |\nabla u(x)|^2 dx &\leq C|R|\delta(R)^{-2}\delta(R)^{-2d}\int_{\Gamma\cap B_R}\int_{\Gamma\cap 100B_R} |g(x) - g(y)|^2 d\sigma(x) d\sigma(y) \\ &\leq C\delta(R)^{n-2d-2}\int_{\Gamma\cap B_R}\int_{\Gamma\cap 100B_R} |g(x) - g(y)|^2 d\sigma(x) d\sigma(y) \end{aligned} \tag{7.20}$$

because $\mathcal{W}(R)$ has at most C elements. We multiply by $\delta(R)^{d+1-n}$, sum over R, and get that

$$\begin{aligned} \|u\|_W^2 &\leq C\sum_{R\in\mathcal{W}} \delta(R)^{-d-1}\int_{\Gamma\cap B_R}\int_{\Gamma\cap 100B_R} |g(x) - g(y)|^2 d\sigma(x) d\sigma(y) \\ &\leq C\int_\Gamma\int_\Gamma |g(x) - g(y)|^2 h(x,y) d\sigma(x) d\sigma(y), \end{aligned} \tag{7.21}$$

where we set

$$h(x,y) = \sum_R \delta(R)^{-d-1}, \tag{7.22}$$

and we sum over $R \in \mathcal{W}$ such that $x \in B_R$ and $y \in 100B_R$. Notice that $|x-y| \leq 101\delta(R)$, so we only sum over R such that $\delta(R) \geq |x-y|/101$.

Let us fix x and y, and evaluate $h(x,r)$. For each scale (each value of $\mathrm{diam}(R)$), there are less than C cubes $R \in \mathcal{W}$ that are possible, because $x \in B_R$ implies that $\mathrm{dist}(x,R) \leq 3\delta(R)$. So the contribution of the cubes for which $\mathrm{diam}(R)$ is of the order r is less than Cr^{-d-1}. We sum over the scales (larger than $C^{-1}|x-y|$) and get less than $C|x-y|^{-d-1}$. That is, $h(x,y) \leq C|x-y|^{-d-1}$ and

$$\|u\|_W^2 \leq C\int_\Gamma\int_\Gamma \frac{|g(x) - g(y)|^2}{|x-y|^{d+1}} d\sigma(x) d\sigma(y) = C\|g\|_H^2, \tag{7.23}$$

as needed for (7.9). Theorem 7.3 follows. □

We end the section with the density in H of (traces of) smooth functions.

LEMMA 7.4. *For every $g \in H$, we can find a sequence $(v_k)_{k\in\mathbb{N}}$ in $C^\infty(\mathbb{R}^n)$ such that Tv_k converges to g in H in $L^1_{loc}(\Gamma,\sigma)$, and σ-a.e. pointwise.*

Notice that since v_k is continuous across Γ, Tv_k is the restriction of v_k to Γ, and we get the density in H of continuous functions on Γ, for the same three convergences.

PROOF. The quickest way to prove this will be to use Theorem 3.4, Theorem 7.3 and the results in Section 5.

Let $g \in H$ be given. Let ρ_ϵ be defined as in Lemma 5.3, and set $v_\varepsilon = \rho_\epsilon * Eg$ and $g_\varepsilon = Tv_\varepsilon$. Theorem 7.3 says that $Eg \in W$; then by Lemma 5.3, $v_\varepsilon = \rho_\epsilon * Eg$ lies in $C^\infty(\mathbb{R}^n) \cap W$. We still need to check that g_ε tends to g for the three types of convergence.

By Lemma 5.3, $v_\varepsilon = \rho_\epsilon * Eg$ converges to Eg in $L^1_{loc}(\mathbb{R}^n)$ and in W, and then (5.16) implies that $g_\epsilon = Tv_\varepsilon$ tends to $g = T(Eg)$ in $L^1_{loc}(\Gamma, \sigma)$.

The convergence in H is the consequence of the bounds

$$\|g - g_\epsilon\|_H \leq \|T(Eg - v_\epsilon)\|_H \leq C\|Eg - v_\epsilon\|_W \tag{7.24}$$

that come from Theorem 7.3, plus the fact that the right-hand side converges to 0 thanks to Lemma 5.3.

For the a.e. pointwise convergence, let us cheat slightly: we know that the g_ε converge to g in $L^1_{loc}(\Gamma, \sigma)$; we can then use the diagonal process to extract a sequence of g_ε that converges pointwise a.e. to g, which is enough for the lemma. □

CHAPTER 8

Definition of Solutions

The aim of the following chapters is to define the harmonic measure on Γ. We follow the presentation of Kenig [**Ken**, Sections 1.1 and 1.2].

In addition to W, we introduce a local version of W. Let $E \subset \mathbb{R}^n$ be an open set. The set of function $W_r(E)$ is defined as

$$W_r(E) = \{f \in L^1_{loc}(E),\ \varphi f \in W \text{ for all } \varphi \in C_0^\infty(E)\} \tag{8.1}$$

where the function φf is seen as a function on $\mathbb{R}^n$ (since φf is compactly supported in E, it can be extended by 0 outside E). The inclusion $W \subset W_r(E)$ is given by Lemma 5.4.

Let us discuss a bit more about our newly defined spaces. First, we claim that

$$W_r(E) \subset \{f \in L^1_{loc}(E),\ \nabla f \in L^2_{loc}(E, w)\}, \tag{8.2}$$

where here ∇f denotes the distributional derivative of f in E. To see this, let $f \in W_r(E)$ be given; we just need to see that $\nabla f \in L^2(K, w)$ for any relatively compact open subset K of E. Pick $\varphi \in C_0^\infty(E)$ such that $\varphi \equiv 1$ on K, and observe that $\varphi f \in W$ by (8.1), so Lemma 3.2 says that φf has a distribution derivative (on $\mathbb{R}^n$) that lies in $L^2(\mathbb{R}^n, w)$. Of course the two distributions ∇f and $\nabla(\varphi f)$ coincide near K, so $\nabla f \in L^2(K, w)$ and our claim follows.

The reverse inclusion $W_r(E) \supset \{f \in L^1_{loc}(E),\ \nabla f \in L^2_{loc}(E, w)\}$ surely holds, but we will not use it. Note that thanks to Lemma 3.2, we do not need to worry, even locally as here, about the difference between having a derivative in $\Omega \cap E$ that lies in $L^2_{loc}(E, w)$ and the apparently stronger condition of having a derivative in E that lies in $L^2_{loc}(E, w)$. Also note that $W_r(\mathbb{R}^n) \neq W$; the difference is that W demands some decay of ∇u at infinity, while $W_r(\mathbb{R}^n)$ doesn't.

LEMMA 8.1. *Let $E \subset \mathbb{R}^n$ be an open set. For every function $u \in W_r(E)$, we can define the trace of u on $\Gamma \cap E$ by*

$$Tu(x) = \lim_{r \to 0} \fint_{B(x,r)} u(z)\, dz \qquad \text{for } \sigma\text{-almost every } x \in \Gamma \cap E, \tag{8.3}$$

and $Tu \in L^1_{loc}(\Gamma \cap E, \sigma)$. Moreover, for every choice of $f \in W_r(E)$ and $\varphi \in C_0^\infty(E)$,

$$T(\varphi u)(x) = \varphi(x) Tu(x) \quad \text{for } \sigma\text{-almost every } x \in \Gamma \cap E. \tag{8.4}$$

PROOF. The existence of $\lim_{r\to 0} \fint_{B(x,r)} u(z)dz$ is easy. If B is any relatively compact ball in E, we can pick $\varphi \in C_0^\infty(E)$ such that $\varphi \equiv 1$ near B. Then $\varphi u \in W$, and the analogue of (8.3) for φu comes with the construction of the trace. This implies the existence of the same limit for f, almost everywhere in $\Gamma \cap B$.

Next we check that $Tu \in L^1_{loc}(\Gamma \cap E, \sigma)$. Let K be a compact set in E; we want to show that $Tu \in L^1(K \cap \Gamma, \sigma)$. Take $\varphi \in C_0^\infty(E)$ such that $\varphi \equiv 1$ on K. Then

$\varphi u \in W$ by definition of $W_r(E)$ and thus

$$\|Tu\|_{L^1(K\cap\Gamma,\sigma)} \leq \|T[\varphi u]\|_{L^1(K\cap\Gamma,\sigma)} \leq C_K\|\varphi u\|_W < +\infty \tag{8.5}$$

by (3.26).

Let us turn to the proof of (8.4). Take $\varphi \in C_0^\infty(E)$ and then choose $\phi \in C_0^\infty(E)$ such that $\phi \equiv 1$ on $\operatorname{supp}\varphi$. According to Lemma 5.4, $T(\varphi\phi u)(x) = \varphi(x)T(\phi u)(x)$ for almost every $x \in \Gamma$. The result then holds by noticing that $\varphi\phi u = \varphi u$ (i.e. $T(\varphi\phi u)(x) = T(\varphi u)(x)$) and $\phi u = u$ on $\operatorname{supp}\varphi$ (i.e. $\varphi(x)T(\phi u)(x) = \varphi(x)T(u)(x)$). □

Let us remind the reader that we will be working with the differential operator $L = -\operatorname{div} A\nabla$, where $A : \Omega \to \mathbb{M}_n(\mathbb{R})$ satisfies, for some constant $C_1 \geq 1$,

- the boundedness condition

$$|A(x)\xi \cdot \nu| \leq C_1 w(x)|\xi|\cdot|\nu| \qquad \forall x \in \Omega,\ \xi,\nu \in \mathbb{R}^n; \tag{8.6}$$

- the ellipticity condition

$$A(x)\xi\cdot\xi \geq C_1^{-1}w(x)|\xi|^2 \qquad \forall x \in \Omega,\ \xi \in \mathbb{R}^n. \tag{8.7}$$

We denote the matrix $w^{-1}A$ by $\mathcal{A}$, so that $\int_\Omega A\nabla u\cdot\nabla v = \int_\Omega \mathcal{A}\nabla u\cdot\nabla v\, dm$. The matrix $\mathcal{A}$ satisfies the unweighted elliptic and boundedness conditions, that is

$$|\mathcal{A}(x)\xi\cdot\nu| \leq C_1|\xi|\cdot|\nu| \qquad \forall x \in \Omega,\ \xi,\nu \in \mathbb{R}^n, \tag{8.8}$$

and

$$\mathcal{A}(x)\xi\cdot\xi \geq C_1^{-1}|\xi|^2 \qquad \forall x \in \Omega,\ \xi \in \mathbb{R}^n. \tag{8.9}$$

Let us introduce now the bilinear form a defined by

$$a(u,v) = \int_\Omega A\nabla u\cdot\nabla v\, dz = \int_\Omega \mathcal{A}\nabla u\cdot\nabla v\, dm. \tag{8.10}$$

From (8.8) and (8.9), we deduce that a is a bounded on $W\times W$ and coercive on W (hence also on W_0). That is,

$$a(u,u) = \int_\Omega \mathcal{A}\nabla u\cdot\nabla u\, dm \geq C_1^{-1}\int_\Omega |\nabla u|^2\, dm = C_1^{-1}\|u\|_W^2 \tag{8.11}$$

for $u \in W$, by (8.9).

DEFINITION 8.2. Let $E \subset \Omega$ be an open set.

We say that $u \in W_r(E)$ is a solution of $Lu = 0$ in E if for any $\varphi \in C_0^\infty(E)$,

$$a(u,\varphi) = \int_\Omega A\nabla u\cdot\nabla\varphi\, dz = \int_\Omega \mathcal{A}\nabla u\cdot\nabla\varphi\, dm = 0. \tag{8.12}$$

We say that $u \in W_r(E)$ is a subsolution (resp. supersolution) in E if for any $\varphi \in C_0^\infty(E)$ such that $\varphi \geq 0$,

$$a(u,\varphi) = \int_\Omega A\nabla u\cdot\nabla\varphi\, dz = \int_\Omega \mathcal{A}\nabla u\cdot\nabla\varphi\, dm \leq 0 \text{ (resp. } \geq 0). \tag{8.13}$$

In particular, subsolutions and supersolutions are always associated to the equation $Lu = 0$. In the same way, each time we say that u is a solution in E, it means that u is in $W_r(E)$ and is a solution of $Lu = 0$ in E.

We start with the following important result, that extends the possible test functions in the definition of solutions.

LEMMA 8.3. *Let $E \subset \Omega$ be an open set and let $u \in W_r(E)$ be a solution of $Lu = 0$ in E. Also denote by E^Γ is the interior of $E \cup \Gamma$. The identity* (8.12) *holds:*

- *when $\varphi \in W_0$ is compactly supported in E;*
- *when $\varphi \in W_0$ is compactly supported in E^Γ and $u \in W_r(E^\Gamma)$;*
- *when $E = \Omega$, $\varphi \in W_0$, and $u \in W$.*

In addition, (8.13) *holds when u is a subsolution (resp. supersolution) in E and φ is a non-negative test function satisfying one of the above conditions.*

REMARK 8.4. The second statement of the Lemma will be used in the following context. Let $B \subset \mathbb{R}^n$ be a ball centered on Γ and let $u \in W_r(B)$ be a solution of $Lu = 0$ in $B \setminus \Gamma$. Then we have

$$a(u, \varphi) = \int_\Omega \mathcal{A}\nabla u \cdot \nabla\varphi \, dm = 0 \tag{8.14}$$

for any $\varphi \in W_0$ compactly supported in B. Similar statements can be written for subsolutions and supersolutions.

PROOF. Let $u \in W_r(E)$ be a solution of $Lu = 0$ on E and let $\varphi \in W_0$ be compactly supported in E. We want to prove that $a(u, \varphi) = 0$.

Let $\widetilde{E}$ be an open set such that $\operatorname{supp} \varphi$ compact in $\widetilde{E}$ and $\widetilde{E}$ is relatively compact in E. By Lemma 5.3, there exists a sequence $(\varphi_k)_{k \geq 1}$ of functions in $C_0^\infty(\widetilde{E})$ such that $\varphi_k \to \varphi$ in W. Observe that the map

$$\phi \to a_{\widetilde{E}}(u, \phi) = \int_{\widetilde{E}} \mathcal{A}\nabla u \cdot \nabla\phi \, dm \tag{8.15}$$

is bounded on W thanks to (8.8) and the fact that $\nabla u \in L^2(\widetilde{E}, w)$ (see (8.2)). Then, since φ and the φ_k are supported in $\widetilde{E}$,

$$a(u, \varphi) = a_{\widetilde{E}}(u, \varphi) = \lim_{k \to +\infty} a_{\widetilde{E}}(u, \varphi_k) = \lim_{k \to +\infty} a(u, \varphi_k) = 0 \tag{8.16}$$

by (8.12).

Now let $u \in W_r(E^\Gamma)$ be a solution of $Lu = 0$ on E and let $\varphi \in W_0$ be compactly supported in E^Γ. We want to prove that $a(u, \varphi) = 0$.

Let $\widetilde{E^\Gamma}$ be an open set such that $\operatorname{supp} \varphi$ is compact in $\widetilde{E^\Gamma}$ and $\widetilde{E^\Gamma}$ is relatively compact in E^Γ. If we look at the proof of Lemma 5.5 (that uses cut-off functions and the smoothing process given by Lemma 5.3), we can see that our $\varphi \in W_0$ can be approached in W by functions $\varphi_k \in C_0^\infty(\widetilde{E^\Gamma} \setminus \Gamma)$. In addition, the map

$$\phi \to a_{\widetilde{E^\Gamma}}(u, \phi) = \int_{\widetilde{E^\Gamma}} \mathcal{A}\nabla u \cdot \nabla\phi \, dm \tag{8.17}$$

is bounded on W thanks to (8.8) and the fact that $\nabla u \in L^2(\widetilde{E^\Gamma}, w)$ (that holds because $u \in W_r(E^\Gamma)$). Then, as before,

$$a(u, \varphi) = a_{\widetilde{E^\Gamma}}(u, \varphi) = \lim_{k \to +\infty} a_{\widetilde{E^\Gamma}}(u, \varphi_k) = \lim_{k \to +\infty} a(u, \varphi_k) = 0. \tag{8.18}$$

The proof of the last point, that is $a(u, \varphi) = 0$ if $u \in W$ and $\varphi \in W_0$, works the same way as before. This time, we use the facts that Lemma 5.3 gives an approximation of φ by functions in $C_0^\infty(\Omega)$ and that $\phi \to a(u, \phi)$ is bounded on W.

Finally, the cases where u is a subsolution or a supersolution have a similar proof. We just need to observe that the smoothing provided by Lemma 5.3 conserves the non-negativity of a test function. □

The first property that we need to know about sub/supersolution is the following stability property.

LEMMA 8.5. *Let $E \subset \Omega$ be an open set.*

- *If $u, v \in W_r(E)$ are subsolutions in E, then $t = \max\{u, v\}$ is also a subsolution in E.*
- *If $u, v \in W_r(E)$ are supersolutions in E, then $t = \min\{u, v\}$ is also a supersolution in E.*

In particular if $k \in \mathbb{R}$, then $(u-k)_+ := \max\{u-k, 0\}$ is a subsolution in E whenever $u \in W_r(E)$ is a subsolution in E and $\min\{u, k\}$ is a supersolution in E whenever $u \in W_r(E)$ is a supersolution in E.

PROOF. It will be enough to prove the the first statement of the lemma, i.e., the fact that $t = \max\{u, v\}$ is a subsolution when u and v are subsolutions. Indeed, the statement about supersolutions will follow at once, because it is easy to see that $u \in W_r(E)$ is a supersolution if and only $-u$ is a subsolution. The remaining assertions are then straightforward consequences of the first ones (because constant functions are solutions).

So we need to prove the first part, and fortunately it will be easy to reduce to the classical situation, where the desired result is proved in [**Sta**, Theorem 3.5]. We need an adaptation, because Stampacchia's proof corresponds to the case where the subsolutions u, v lie in W, and also we want to localize to a place where w is bounded from above and below.

Let $F \subset E$ be any open set with a smooth boundary and a finite number of connected components, and whose closure is compact in E. We define a set of functions W^F as

$$W^F = \{f \in L^1_{loc}(F),\ \nabla f \in L^2(F, w)\}. \tag{8.19}$$

Let us record a few properties of W^F. Since F is relatively compact in $E \subset \Omega$, the weight w is bounded from above and below by a positive constant. Hence W^F is the collection of functions in $L^1_{loc}(F)$ whose distributional derivative lies in $L^2(F)$. Since F is bounded and has a smooth boundary, these functions lie in $L^2(F)$ (see [**Maz**, Corollary 1.1.11]). Of course Mazya states this when F is connected, but we here F has a finite number of components, and we can apply the result to each one. So W^F is the 'classical' (where the weight is plain) Sobolev space on F. That is,

$$W^F = \{f \in L^2(F),\ \nabla f \in L^2(F)\}. \tag{8.20}$$

Notice that u and v lie in W^F, so they are "classical" subsolutions of L in F, where (since F is relatively compact in $E \subset \Omega$) w is bounded from above and below, and hence L is a classical elliptic operator. Then, by [**Sta**, Theorem 3.5], $t = \max\{u, v\}$ is also a classical subsolution in F. This means that $a(t, \varphi) \leq 0$ for $\varphi \in C_0^\infty(F)$.

Now we wanted to prove this for every $\varphi \in C_0^\infty(E)$, and it is enough to observe that if $\varphi \in C_0^\infty(E)$ is given, then we can find an open set $F \subset\subset E$ that contains the support of φ, and with the regularity properties above. Hence t is a subsolution in E, and the lemma follows. It was fortunate for this argument that the notion of subsolution does not come with precise estimates that would depend on w. □

In the sequel, the notation sup and inf are used for the essential supremum and essential infimum, since they are the only definitions that makes sense for the functions in W or in $W_r(E)$, $E \subset \mathbb{R}^n$ open. Also, when we talk about solutions or subsolutions and don't specify, this will always refer to our fixed operator L. We now state some classical regularity results inside the domain.

LEMMA 8.6 (interior Caccioppoli inequality). *Let $E \subset \Omega$ be an open set, and let $u \in W_r(E)$ be a non-negative subsolution in E. Then for any $\alpha \in C_0^\infty(E)$,*

$$\int_\Omega \alpha^2 |\nabla u|^2 dm \leq C \int_\Omega |\nabla \alpha|^2 u^2 dm, \tag{8.21}$$

where C depends only upon the dimensions n and d and the constant C_1.

In particular, if B is a ball of radius r such that $2B \subset \Omega$ and $u \in W_r(2B)$ is a non-negative subsolution in $2B$, then

$$\int_B |\nabla u|^2 dm \leq C r^{-2} \int_{2B} u^2 dm. \tag{8.22}$$

PROOF. Let $\alpha \in C_0^\infty(E)$. We set $\varphi = \alpha^2 u$. Since $u \in W_r(E)$, the definition yields $\varphi \in W$. Moreover φ is compactly supported in E (and in particular $\varphi \in W_0$). The first item of Lemma 8.3 yields

$$\int_\Omega \mathcal{A}\nabla u \cdot \nabla \varphi \, dm \leq 0. \tag{8.23}$$

By the product rule, $\nabla \varphi = \alpha^2 \nabla u + 2\alpha u \nabla \alpha$. Thus (8.23) becomes

$$\int_\Omega \alpha^2 \mathcal{A}\nabla u \cdot \nabla u \, dm \leq -2 \int_\Omega \alpha u \, \mathcal{A}\nabla u \cdot \nabla \alpha \, dm. \tag{8.24}$$

It follows from this and the ellipticity and boundedness conditions (8.9) and (8.8) that

$$\int_\Omega \alpha^2 |\nabla u|^2 dm \leq C \int_\Omega |\alpha||\nabla u||u||\nabla \alpha| \, dm \tag{8.25}$$

and then

$$\int_\Omega \alpha^2 |\nabla u|^2 dm \leq C \left(\int_\Omega \alpha^2 |\nabla u|^2 dm \right)^{\frac{1}{2}} \left(\int_\Omega u^2 |\nabla \alpha|^2 dm \right)^{\frac{1}{2}} \tag{8.26}$$

by the Cauchy-Schwarz inequality. Consequently,

$$\int_\Omega \alpha^2 |\nabla u|^2 dm \leq C \int_\Omega |\nabla \alpha|^2 u^2 dm, \tag{8.27}$$

which is (8.22). Lemma 8.11 follows since (8.37) is a straightforward application of (8.22) when $E = 2B$, $\alpha \equiv 1$ on B and $|\nabla \alpha| \leq \frac{2}{r}$. □

LEMMA 8.7 (interior Moser estimate). *Let $p > 0$ and B be a ball such that $3B \subset \Omega$. If $u \in W_r(3B)$ is a non-negative subsolution in $2B$, then*

$$\sup_B u \leq C \left(\frac{1}{m(2B)} \int_{2B} u^p \, dm \right)^{\frac{1}{p}}, \tag{8.28}$$

where C depends on n, d, C_1 and p.

PROOF. For this lemma and the next ones, we shall use the fact that since $2B$ is far from Γ, our weight w is under control there, and we can easily reduce to the classical case. Let x and r denote the center and the radius of B. Since $3B \subset \Omega$, $\delta(x) \geq 3r$. For any $z \in 2B$, $\delta(x) - 2r \leq \delta(z) \leq \delta(x + 2r)$, hence

$$\frac{1}{3} \leq 1 - \frac{2r}{\delta(x)} \leq \frac{\delta(z)}{\delta(x)} \leq 1 + \frac{2r}{\delta(x)} \leq \frac{5}{3} \tag{8.29}$$

and consequently

$$C_{n,d}^{-1} w(x) \leq w(z) \leq C_{n,d}\, w(x). \tag{8.30}$$

Let $u \in W_r(3B)$ be a non-negative subsolution in $2B$. Thanks to (8.2) and (8.30), the gradient ∇u lies in $L^2(2B)$. By the Poincaré's inequality, $u \in L^2(2B)$ and thus u lies in the classical (with no weight) Sobolev space W^{2B} of (8.20).

Consider the differential operator $\widetilde{L} = -\operatorname{div} \widetilde{A}\nabla$ with $\widetilde{A}(z) = \mathcal{A}(z)\frac{w(z)}{w(x)}$. Thanks to (8.30), (8.8) and (8.9), $\widetilde{A}(z)$ satisfies the elliptic condition and the boundedness condition (8.8) and (8.9), in the domain $2B$, and with the constant $C_{n,d}C_1$. The condition satisfied by a subsolution (of $Lu = 0$) on $2B$ can be rewritten

$$\int_{2B} \widetilde{A}\nabla u \cdot \nabla \varphi \leq 0, \tag{8.31}$$

and so we are back in the situation of the classical elliptic case. By [**Ken**, Lemma 1.1.8], for instance,

$$\sup_B u \leq C \left(\fint_{2B} u^p(z)\, dz \right)^{\frac{1}{p}}, \tag{8.32}$$

and (8.28) follows from this and (2.13) □

LEMMA 8.8 (interior Hölder continuity). *Let $x \in \Omega$ and $R > 0$ be such that $B(x, 3R) \subset \Omega$, and let $u \in W_r(B(x, 3R))$ be a solution in $B(x, 2R)$. Write $\operatorname{osc}_B u$ for $\sup_B u - \inf_B u$. Then there exists $\alpha \in (0, 1]$ and $C > 0$ such that for any $0 < r < R$,*

$$\operatorname*{osc}_{B(x,r)} u \leq C \left(\frac{r}{R}\right)^{\alpha} \left(\frac{1}{m(B(x,R))} \int_{B(x,R)} u^2\, dm \right)^{\frac{1}{2}}, \tag{8.33}$$

where α and C depend only on n, d, and C_1. Hence u is (possibly after modifying it on a set of measure 0) locally Hölder continuous with exponent α.

PROOF. This lemma and the next one follow from the classical results (see for instance [**Ken**, Section 1.1], or [**GT**, Sections 8.6, 8.8 and 8.9]), by the same trick as for Lemma 8.7: we observe that L is a constant times a classical elliptic operator on $2B$. □

LEMMA 8.9 (Harnack). *Let B be a ball such that $3B \subset \Omega$, and let $u \in W_r(3B)$ be a non-negative solution in $3B$. Then*

$$\sup_B u \leq C \inf_B u, \tag{8.34}$$

where C depends only on n, d and C_1.

For the next lemma, we shall need the Harnack chains from Lemma 2.1.

LEMMA 8.10. *Let K be a compact set of Ω and let $u \in W_r(\Omega)$ be a non-negative solution in Ω. Then*

$$\sup_K u \leq C_K \inf_K u, \tag{8.35}$$

where C_K depends only on n, d, C_0, C_1, $\mathrm{dist}(K,\Gamma)$ *and* $\mathrm{diam}\, K$.

PROOF. Let K be a compact set in Ω. We can find $r > 0$ and $k \geq 1$ such that $\mathrm{dist}(K,\Gamma) \geq r$ and $\mathrm{diam}\, K \leq kr$. Now let $x, y \in K$ be given. Notice that $\delta(x) \geq r$, $\delta(y) \geq r$ and $|x-y| \leq kr$, so Lemma 2.1 implies the existence of a path of length at most by $(k+1)r$ that joins x to y and stays at a distance larger than some ϵ (that depends on C_0, d, n, r and k) of Γ. That is, we can find a finite collection of balls $B_1, \dots, B_n$ (n bounded uniformly on $x, y \in K$) such that $3B_i \subset \Omega$, B_1 is centered on x, B_n is centered on y, and $B_i \cap B_{i+1} \neq \emptyset$. It remains to use n times Lemma 8.9 to get that

$$u(x) \leq C^n u(y) \leq C_K u(y). \tag{8.36}$$

Lemma 8.10 follows. □

We also need analogues at the boundary of the previous results. For these we cannot immediately reduce to the classical case, but we will be able to copy the proofs. Of course we shall use our trace operator to define boundary conditions, say, in a ball B, and this is the reason why we want to use the space is $W_r(B)$ defined by (8.1). We cannot use $W_r(B \setminus \Gamma)$ instead, because we need some control on u near Γ to define $T(u)$.

In the sequel, we will use the expression '$Tu = 0$ a.e. on B', for a function $u \in W_r(B)$, to mean that Tu, which is defined on $\Gamma \cap B$ and lies in $L^1_{loc}(B \cap \Gamma, \sigma)$ thanks to Lemma 8.1, is equal to 0 σ-almost everywhere on $\Gamma \cap B$. The expression '$Tu \geq 0$ a.e. on B' is defined similarly.

We start with the Caccioppoli inequality on the boundary.

LEMMA 8.11 (Caccioppoli inequality on the boundary). *Let $B \subset \mathbb{R}^n$ be a ball of radius r centered on Γ, and let $u \in W_r(2B)$ be a non-negative subsolution in $2B \setminus \Gamma$ such that $T(u) = 0$ a.e. on $2B$. Then for any $\alpha \in C_0^\infty(2B)$,*

$$\int_{2B} \alpha^2 |\nabla u|^2 dm \leq C \int_{2B} |\nabla \alpha|^2 u^2 dm, \tag{8.37}$$

where C depends only on the dimensions n and d and the constant C_1. In particular, we can take $\alpha \equiv 1$ on B and $|\nabla \alpha| \leq \frac{2}{r}$, which gives

$$\int_B |\nabla u|^2 dm \leq C r^{-2} \int_{2B} u^2 dm. \tag{8.38}$$

PROOF. We can proceed exactly as for Lemma 8.6, except that the initial estimate (8.23) needs to be justified differently. Here we choose to apply the second item of Lemma 8.3, as explained in Remark 8.4. That is, $E = 2B \setminus \Gamma$ and $E^\Gamma = 2B$.

So we check the assumptions. We set, as before, $\varphi = \alpha^2 u$. First observe that $\varphi \in W$ because $u \in W_r(2B)$ and $\alpha \in C_0^\infty(2B)$. Moreover, $\varphi \in W_0$ because, if we let $\phi \in C_0^\infty(2B)$ be such that $\phi \equiv 1$ on a neighborhood of $\mathrm{supp}\, \alpha$, Lemma 5.4 says that $T(\varphi) = T(\alpha^2 \phi u) = \alpha^2 T(\phi u) = 0$ a.e. on Γ. In addition, φ is compactly supported in $2B$ because α is, and $u \in W_r(2B)$ by assumption.

Thus φ is a valid test function, Lemma 8.3 applies, (8.23) holds, and the rest of the proof is the same as for Lemma 8.6. □

LEMMA 8.12 (Moser estimates on the boundary). *Let B be a ball centered on Γ. Let $u \in W_r(2B)$ be a non-negative subsolution in $2B \setminus \Gamma$ such that $Tu = 0$ a.e. on $2B$. Then*

$$\sup_B u \leq C \left(m(2B)^{-1} \int_{2B} u^2 dm \right)^{\frac{1}{2}}, \tag{8.39}$$

where C depends only on the dimensions d and n and the constants C_0 and C_1.

PROOF. This proof will be a little longer, but we will follow the ideas used by Stampacchia in [**Sta**, Section 5]. The aim is to use the so-called Moser iterations. We start with some consequences of Lemma 8.11.

Pick $2^* \in (2, +\infty)$ in the range of p satisfying the Sobolev-Poincaré inequality (4.31); for instance take $2^* = \frac{2n}{n-1}$. Let u be as in the statement and let $B = B(x,r)$ be a ball centered on Γ. We claim that for any $\alpha \in C_0^\infty(2B)$,

$$\int_{2B} (\alpha u)^2 dm \leq C r^2 m(\operatorname{supp} \alpha u)^{1-\frac{2}{2^*}} m(2B)^{\frac{2}{2^*}-1} \int_{2B} |\nabla \alpha|^2 u^2 dm \tag{8.40}$$

where in fact we abuse notation and set $\operatorname{supp} \alpha u = \{\alpha u > 0\}$. Indeed, by Hölder's inequality and the Sobolev-Poincaré inequality (4.31),

$$\begin{aligned} \int_{\mathbb{R}^n} (\alpha u)^2 dm &\leq C m(\operatorname{supp} \alpha u)^{1-\frac{2}{2^*}} \left(\int_{2B} (\alpha u)^{2^*} dm \right)^{\frac{2}{2^*}} \\ &\leq C r^2 m(\operatorname{supp} \alpha u)^{1-\frac{2}{2^*}} m(2B)^{\frac{2}{2^*}-1} \int_{2B} |\nabla[\alpha u]|^2 dm. \end{aligned} \tag{8.41}$$

The last integral can be estimated, using Caccioppoli's inequality (Lemma 8.11), by

$$\begin{aligned} \int_{2B} |\nabla(\alpha u)|^2 dm &\leq 2 \int_{2B} |\nabla \alpha|^2 u^2 dm + 2 \int_{2B} |\nabla u|^2 \alpha^2 dm \\ &\leq C \int_{2B} |\nabla \alpha|^2 u^2 dm. \end{aligned} \tag{8.42}$$

Our claim claim (8.40) follows.

Recall that $B = B(x,r)$, with $x \in \Gamma$. Since u is a subsolution in $2B \setminus \Gamma$, Lemma 8.5 says that $(u-k)_+ := \max\{u-k, 0\}$ is a non-negative subsolution in $2B \setminus \Gamma$. For any $0 < s < t \leq 2r$, we choose a smooth function α supported in $B(x,t)$, such that $0 \leq \alpha \leq 1$, $\alpha \equiv 1$ on $B(x,s)$, and $|\nabla \alpha| \leq \frac{2}{t-s}$. By (8.40) (applied to $(u-k)_+$ and this function α),

$$\int_{A(k,s)} |u-k|^2 dm \leq C \frac{r^2}{(t-s)^2} m(A(k,t))^{1-\frac{2}{2^*}} m(2B)^{\frac{2}{2^*}-1} \int_{A(k,t)} |u-k|^2 dm \tag{8.43}$$

where $A(k,s) = \{y \in B(x,s),\, u(y) > k\}$. If $h > k$, we have also,

$$(h-k)^2 m(A(h,s)) \leq \int_{A(h,s)} |u-k|^2 dm \leq \int_{A(k,s)} |u-k|^2 dm. \tag{8.44}$$

Define

$$a(h,s) = m(A(h,s)) \tag{8.45}$$

and

$$u(h,s) = \int_{A(h,s)} |u-h|^2 dm; \tag{8.46}$$

thus

$$\begin{cases} u(k,s) \le \dfrac{Cr^2 m(2B)^{\frac{2}{2^*}-1}}{(t-s)^2}\, u(k,t)[a(k,t)]^{1-\frac{2}{2^*}} \\ a(h,s) \le \dfrac{1}{(h-k)^2}\, u(k,t) \end{cases} \tag{8.47}$$

or, if we set $\kappa = 1 - \frac{2}{2^*} > 0$,

$$\begin{cases} u(k,s) \le \dfrac{Cr^2 m(2B)^{-\kappa}}{(t-s)^2}\, u(k,t)[a(k,t)]^{\kappa} \\ a(h,s) \le \dfrac{1}{(h-k)^2}\, u(k,t). \end{cases} \tag{8.48}$$

Notice also that $u(h,s) \le u(k,s)$ because $A(h,s) \subset A(k,s)$ and $|u-h|^2 \le |u-k|^2$ on $A(h,s)$.

Let $\epsilon > 0$ be given, to be chosen later. The estimates (8.48) yield

$$u(h,s)^{\epsilon} a(h,s) \le u(k,s)^{\epsilon} a(h,s) \le \frac{Cr^{2\epsilon} m(2B)^{-\epsilon\kappa}}{(t-s)^{2\epsilon}(h-k)^2}\, u(k,t)^{\epsilon+1} a(k,t)^{\epsilon\kappa}. \tag{8.49}$$

Following [**Sta**], we define a function of two variables φ by

$$\varphi(h,s) = u(h,s)^{\epsilon} a(h,s) \quad \text{for } h > 0 \text{ and } 0 < s < 2r. \tag{8.50}$$

Notice that $\varphi(h,s) \ge 0$. When s is fixed, $\varphi(h,s)$ is non increasing in h, and when h is fixed, $\varphi(h,s)$ is non decreasing in s. We want to show that

$$\varphi(h,s) \le \frac{K}{(h-k)^{\alpha}(t-s)^{\gamma}}\, [\varphi(k,t)]^{\beta} \tag{8.51}$$

for some choice of positive constants K, α and γ, and some $\beta > 1$, because if we do so we shall be able to use Lemma 5.1 in [**Sta**] directly.

It is a good idea to choose ϵ so that

$$\begin{cases} \beta\epsilon = \epsilon + 1, \\ \quad \beta = \epsilon\kappa. \end{cases} \tag{8.52}$$

for some $\beta > 1$. Choose $\beta = \frac{1}{2} + \sqrt{\frac{1}{4} + \kappa} > 1$ and $\epsilon = \frac{\beta}{\kappa} > 0$. An easy computation proves that (ϵ, β) satisfies (8.52). With this choice, (8.49) becomes

$$\varphi(h,s) \le \frac{Cr^{2\epsilon} m(2B)^{-\epsilon\kappa}}{(t-s)^{2\epsilon}(h-k)^2} \varphi(k,t)^{\beta}, \tag{8.53}$$

which is exactly (8.51) with $K = Cr^{2\epsilon} m(2B)^{-\epsilon\kappa}$, $\alpha = 2$ and $\gamma = 2\epsilon$.

So we can apply Lemma 5.1 in [**Sta**], which says that

$$\varphi(\mathfrak{d}, r) = 0, \tag{8.54}$$

where $\mathfrak{d}$ is given by

$$\mathfrak{d}^{\alpha} = \frac{2^{\beta\frac{\alpha+\beta}{\beta-1}} K[\varphi(0,2r)]^{\beta-1}}{r^{\gamma}}. \tag{8.55}$$

We replace and get that we can take

$$\mathfrak{d}^2 = Cr^{2\epsilon} m(2B)^{-\epsilon\kappa}\, \frac{\varphi(0,2r)^{\beta-1}}{r^{\gamma}} = Cm(2B)^{-\epsilon\kappa} \varphi(0,2r)^{\beta-1}. \tag{8.56}$$

Notice that $\varphi(\mathfrak{d},r)=0$ implies that $a(\mathfrak{d},r)=0$, which in turn implies that $u\le\mathfrak{d}$ a.e. on $B=B(x,r)$. Moreover, by definition of a, we have $a(0,2r)\le m(2B)$. Thus

$$\begin{aligned}\sup_{B(x,r)} u &\le \mathfrak{d} \le Cm(2B)^{-\epsilon\kappa/2}u(0,2r)^{(\beta-1)\varepsilon/2}a(0,2r)^{(\beta-1)/2}\\ &\le Cu(0,2r)^{\epsilon(\beta-1)/2}m(2B)^{(\beta-1-\epsilon\kappa)/2}.\end{aligned}\tag{8.57}$$

The first line in (8.52) yields $\epsilon(\beta-1)=1$ and the second line in (8.52) yields $\beta-1-\epsilon\kappa=-1$. Besides, $u(0,2r)=\int_{2B}u^2dm$ because u is nonnegative. Hence

$$\sup_B u\le C\left(m(2B)^{-1}\int_{2B}u^2dm\right)^{\frac12},\tag{8.58}$$

which is the desired conclusion. □

LEMMA 8.13 (Moser estimate at the boundary for general p). *Let $p>0$. Let B be a ball centered on Γ. Let $u\in W_r(2B)$ be a non-negative subsolution in $2B\setminus\Gamma$ such that $Tu=0$ a.e. on $2B$. Then*

$$\sup_B u\le C_p\left(m(2B)^{-1}\int_{2B}u^pdm\right)^{\frac1p},\tag{8.59}$$

where C_p depends only on the dimensions n and d, the constants C_0 and C_1, and the exponent p.

PROOF. Lemma 8.13 can be deduced from Lemma 8.12 by a simple iterative argument. The proof is fairly similar to the very end of the proof of [**HL**, Chapter IV, Theorem 1.1]. Nevertheless, because the proof in [**HL**] doesn't hold at the boundary (and for the sake of completeness), we give a proof here.

First, let us prove that we can improve (8.39) into the following: if B is a ball centered on Γ and $u\in W_r(B)$ is a non-negative subsolution on $B\cap\Omega$ such that $Tu=0$ a.e. on B, then for any $\theta\in(0,1)$ (in practice, close to 1),

$$\sup_{\theta B} u\le C(1-\theta)^{-\frac n2}\left(m(B)^{-1}\int_B u^2\,dm\right)^{\frac12},\tag{8.60}$$

where $C>0$ depends only on n, d, C_0 and C_1.

Let B be a ball centered on Γ, with radius r, and let $\theta\in(0,1)$. Choose $x\in\theta B$. Two cases may happen: either $\delta(x)\ge\frac{1-\theta}{6}r$ or $\delta(x)<\frac{1-\theta}{6}r$. In the first case, if $\delta(x)\ge\frac{1-\theta}{6}r$, we apply Lemma 8.7 to the ball $B(x,\frac{1-\theta}{20}r)$ (notice that $B(x,\frac{1-\theta}{10}r)\subset B\cap\Omega$). We get that

$$\begin{aligned}u(x)&\le C\left(\frac{1}{m(B(x,\frac{1-\theta}{10}r))}\int_{B(x,\frac{1-\theta}{10}r)}u^2\,dm\right)^{\frac12}\\ &\le C\left(\frac{m(B(x,2r))}{m(B(x,\frac{1-\theta}{10}r)}\right)^{\frac12}\left(\frac{1}{m(B)}\int_B u^2\,dm\right)^{\frac12}\\ &\le C(1-\theta)^{-\frac n2}\left(m(B)^{-1}\int_B u^2\,dm\right)^{\frac12}\end{aligned}\tag{8.61}$$

by (2.8). In the second case, when $\delta(x)\le\frac{1-\theta}{6}r$, we take $y\in\Gamma$ such that $|x-y|=\delta(x)$. Remark that $y\in\frac{1+\theta}{2}B$ and then $B(y,\frac{1-\theta}{2}r)\subset B$. We apply then Lemma

8.12 to the ball $B(y, \frac{1-\theta}{6}r)$ in order to get

$$
\begin{aligned}
u(x) \leq \sup_{B(y,\frac{1-\theta}{6}r)} u &\leq C\left(\frac{1}{m(B(x,\frac{1-\theta}{3}r))}\int_{B(x,\frac{1-\theta}{3}r)} u^2\,dm\right)^{\frac{1}{2}} \\
&\leq C\left(\frac{m(B(x,2r))}{m(B(x,\frac{1-\theta}{3}r)}\right)^{\frac{1}{2}}\left(\frac{1}{m(B)}\int_B u^2\,dm\right)^{\frac{1}{2}} \\
&\leq C(1-\theta)^{-\frac{n}{2}}\left(m(B)^{-1}\int_B u^2\,dm\right)^{\frac{1}{2}}
\end{aligned}
\tag{8.62}
$$

with (2.8). The claim (8.60) follows.

Let us prove now (8.59). Without loss of generality, we can restrict to the case $p < 2$, since the case $p \geq 2$ can be deduced from Lemma 8.12 and Hölder's inequality.

Let $B = B(x,r)$ be a ball and let $u \in W_r(2B)$ be a non-negative subsolution on $2B \setminus \Gamma$ such that $Tu = 0$ on $2B$. Set for $i \in \mathbb{N}$,

$$
r_i := r\sum_{j=0}^{i} 3^{-j} = \frac{3}{2}r(1-3^{-i-1}) < \frac{3}{2}r.
$$

Note that $\frac{r_i}{r_i - r_{i-1}} = \frac{3^{i+1}-1}{2} \leq 3^{i+1}$. As a consequence, for any $i \in \mathbb{N}^*$, (8.60) yields

$$
\begin{aligned}
\sup_{B(x,r_{i-1})} u &\leq C3^{\frac{in}{2}}\left(\frac{1}{m(B(x,r_i))}\int_{B(x,r_i)} |u|^2 dm\right)^{\frac{1}{2}} \\
&\leq C3^{\frac{in}{2}}\left(\sup_{B(x,r_i)} u\right)^{1-\frac{p}{2}}\left(\frac{1}{m(B(x,r_i))}\int_{B(x,r_i)} |u|^p dm\right)^{\frac{1}{2}} \\
&\leq C3^{\frac{in}{2}}\left(\sup_{B(x,r_i)} u\right)^{1-\frac{p}{2}}\left(\frac{1}{m(2B)}\int_{B(x,r_i)} |u|^p dm\right)^{\frac{1}{2}}.
\end{aligned}
\tag{8.63}
$$

Set $\alpha = 1 - \frac{p}{2}$. By taking the power α^{i-1} of the inequality (8.63), where i is a positive integer, we obtain

(8.64)
$$
\left(\sup_{B(x,r_{i-1})} u\right)^{\alpha^{i-1}} \leq C^{\alpha^{i-1}}(3^{\frac{in}{2}})^{\alpha^{i-1}}\left(\sup_{B(x,r_i)} u\right)^{\alpha^i}\left(m(2B)^{-1}\int_{B(x,r_i)} |u|^p dm\right)^{\frac{1}{2}\alpha^i},
$$

where C is independent of i (and also p, x, r and u). An immediate induction gives, for any $i \geq 1$,

(8.65)
$$
\sup_{B(x,r)} u \leq C^{\sum_{j=0}^{i-1}\alpha^j}\left(\prod_{j=1}^{i} 3^{\frac{jn}{2}\alpha^{j-1}}\right)\left(\sup_{B(x,r_i)} u\right)^{\alpha^i}\left(m(2B)^{-1}\int_{B(x,r_i)} |u|^p dm\right)^{\frac{1}{2}\sum_{j=0}^{i-1}\alpha^j},
$$

and if we apply Corollary 8.39 once more, we get that

$$\sup_B u \leq C^{\sum_{j=0}^{i} \alpha^j} \left(\prod_{j=1}^{i+1} 3^{\frac{jn}{2}\alpha^{j-1}} \right) \left(m(2B)^{-1} \int_{2B} |u|^p dm \right)^{\frac{1}{2}\sum_{j=0}^{i-1} \alpha^j}$$

$$\times \left(m(2B)^{-1} \int_{\frac{3}{2}B} |u|^2 dm \right)^{\frac{\alpha^i}{2}}. \tag{8.66}$$

We want to to take the limit when i goes to $+\infty$. Since $u \in W_r(2B)$, the quantity $\int_{\frac{3}{2}B} |u|^2 dm$ is finite and thus

$$\lim_{i \to +\infty} \left(m(2B)^{-1} \int_{\frac{3}{2}B} |u|^2 dm \right)^{\frac{\alpha^i}{2}} = 1 \tag{8.67}$$

because we took p such that $\alpha = 1 - \frac{p}{2} < 1$. Note also that

$$\lim_{i \to +\infty} \sum_{j=0}^{i-1} \alpha^j = \frac{2}{p} \quad \text{and} \quad \lim_{i \to +\infty} \frac{1}{2} \sum_{j=0}^{i-1} \alpha^j = \frac{1}{p}. \tag{8.68}$$

Furthermore, $\prod_{j=1}^{i+1} 3^{\frac{jn}{2}\alpha^{j-1}}$ has a limit (that depends on p and n) when $j < +\infty$ because,

$$\sum_{j=1}^{\infty} \frac{jn}{2} \alpha^{j-1} = \frac{n}{2} \sum_{j=1}^{+\infty} j\alpha^{j-1} < +\infty. \tag{8.69}$$

These three facts prove that the limit when $i \to +\infty$ of the right-hand side of (8.66) exists and

$$\sup_B u \leq C_p \left(m(2B)^{-1} \int_{2B} |u|^p dm \right)^{\frac{1}{p}}, \tag{8.70}$$

which is the desired result. □

Next comes the Hölder continuity of the solutions at the boundary. We start with a boundary version of the density property.

LEMMA 8.14. *Let B be a ball centered on Γ and $u \in W_r(4B)$ be a non-negative supersolution in $4B \setminus \Gamma$ such that $Tu = 1$ a.e. on $4B$. Then*

$$\inf_B u \geq C^{-1}, \tag{8.71}$$

where $C > 0$ depends only on the dimensions d, n and the constants C_0, C_1.

PROOF. The ideas of the proof are taken from the Density Theorem (Section 4.3, Theorem 4.9) in [**HL**]. The result in [**HL**] states, roughly speaking, that (8.71) holds whenever u is a supersolution in $4B \subset \Omega$ such that $u \geq 1$ on a large piece of B; and its proof relies on a Poincaré inequality on balls for functions that equal 0 on a big piece of the considered ball.

We will adapt this argument to the case where B is centered on Γ and we will rely on the Poincaré inequality given by Lemma 4.1.

Let B and u be as in the statement. Let $\delta \in (0,1)$ be small (it will be used to avoid some functions to take the value 0) and set $u_\delta = \min\{1, u + \delta\}$ and

$v_\delta := -\Phi_\delta(u_\delta)$, where Φ is a smooth Lipschitz function defined on $\mathbb{R}$ such that $\Phi_\delta(s) = -\ln(s)$ when $s \in [\delta, 1]$.

The plan of the proof is: first we prove that v_δ is a subsolution, and then we use the Moser estimate and the Poincaré inequality given Lemma 8.12 and 4.1 respectively. It will give that the supremum of v_δ on B is bounded by the L^2-norm of the gradient of v_δ. Then, we will test the supersolution u_δ against an appropriate test function, which will give that the $L^2(2B)$ bound on ∇v_δ - and thus the supremum of v_δ on B - can be bounded by a constant independent of δ. This will yield a lower bound on $u_\delta(x)$ which is uniform in δ and $x \in B$.

So we start by proving that

$$v_\delta \in W_r(4B) \text{ is a subsolution in } 4B \setminus \Gamma \text{ such that } Tv_\delta = 0 \text{ a.e. on } 4B. \tag{8.72}$$

Let $\varphi \in C_0^\infty(\Omega \cap 4B)$. Choose $\phi \in C_0^\infty(\Omega \cap 4B)$ such that $\phi \equiv 1$ on $\operatorname{supp}\varphi$. Then for $y \in \Omega$,

$$v_\delta(y)\varphi(y) = \Phi_\delta(\min\{1, (u(y)+\delta)\phi(y)\})\varphi(y). \tag{8.73}$$

Since $u \in W_r(4B)$, it follows that $u\phi \in W$ and thus $(u+\delta)\phi \in W$. Consequently, we obtain $\min\{1, (u+\delta)\phi\} \in W$ by Lemma 6.1 (b), then $\Phi_\delta(\min\{1, (u+\delta)\phi\}) \in W$ by Lemma 6.1 (a) and finally $v_\delta\varphi \in W$ thanks to Lemma 5.4. Hence $v_\delta \in W_r(4B)$. Using the fact that the trace is local and Lemmata 6.1 and 8.1, it is clear that

$$Tv_\delta = -\ln(\min\{1, T(u\phi)+\delta\}) = 0 \quad \text{a.e. on } 4B. \tag{8.74}$$

The claim (8.72) will be proven if we can show that v_δ is a subsolution in $4B \setminus \Gamma$. Let $\varphi \in C_0^\infty(4B \setminus \Gamma)$ be a non-negative function. We have

$$\begin{aligned}\int_\Omega \mathcal{A}\nabla v_\delta \cdot \nabla\varphi \, dm &= -\int_\Omega \frac{\mathcal{A}\nabla u_\delta}{u_\delta} \cdot \nabla\varphi \, dm \\ &= -\int_\Omega \mathcal{A}\nabla u_\delta \cdot \nabla\left(\frac{\varphi}{u_\delta}\right) dm - \int_{4B} \frac{\mathcal{A}\nabla u_\delta \cdot \nabla u_\delta}{u_\delta^2}\varphi \, dm.\end{aligned} \tag{8.75}$$

The second term in the right-hand side is non-positive by the ellipticity condition (8.9). So v_δ is a subsolution if we can establish that

$$\int_\Omega \mathcal{A}\nabla u_\delta \cdot \nabla\left(\frac{\varphi}{u_\delta}\right) dm \geq 0. \tag{8.76}$$

Yet u_δ is a supersolution according to Lemma 8.5. Moreover φ/u_δ is compactly supported in $4B\setminus\Gamma$ and, since $u_\delta \geq \delta > 0$, we deduce from Lemma 6.1 that $\frac{\varphi}{u_\delta} \in W$. So (8.76) is just a consequence of Lemma 8.3. The claim (8.72) follows.

The function v_δ satisfies now all the assumptions of Lemma 8.12 and thus

$$\sup_B v_\delta \leq C\left(m(2B)^{-1}\int_{2B} |v_\delta|^2 dm\right)^{\frac{1}{2}}. \tag{8.77}$$

Since $Tv_\delta = 0$ a.e. on $2B$, the right-hand side can be bounded with the help of (4.13), which gives

$$\sup_B v_\delta \leq Cr\left(m(2B)^{-1}\int_{2B} |\nabla v_\delta|^2 dm\right)^{\frac{1}{2}}. \tag{8.78}$$

We will prove that the right-hand side of (8.78) is bounded uniformly in δ. Use the test function $\varphi = \alpha^2\big(\frac{1}{u_\delta} - 1\big)$ with $\alpha \in C_0^\infty(4B)$, $0 \leq \alpha \leq 1$, $\alpha \equiv 1$ on $2B$ and

$\nabla\alpha \leq \frac{1}{r}$. Note that φ is a non-negative function compactly supported in $4B$ and, by Lemma 6.1, φ is in W and has zero trace, that is $\varphi \in W_0$.

Since u is a supersolution, u_δ is also a supersolution. We test u_δ against φ (this is allowed, thanks to Lemma 8.3) and we get

$$\begin{aligned} 0 \;&\leq\; \int_{\mathbb{R}^n} \mathcal{A}\nabla u_\delta \cdot \nabla\Big[\alpha^2\Big(\frac{1}{u_\delta} - 1\Big)\Big] dm \\ &= \; -\int_{\mathbb{R}^n} \alpha^2 \frac{\mathcal{A}\nabla u_\delta \cdot \nabla u_\delta}{u_\delta^2} dm + 2\int_{\mathbb{R}^n} \alpha\,(1-u_\delta)\,\frac{\mathcal{A}\nabla u_\delta \cdot \nabla\alpha}{u_\delta}\, dm, \end{aligned} \tag{8.79}$$

hence, by the ellipticity and the boundedness of A (see (8.8) and (8.9)),

$$\begin{aligned} \int_{\mathbb{R}^n} \alpha^2 \frac{|\nabla u_\delta|^2}{u_\delta^2} dm &\leq C \int_{\mathbb{R}^n} \alpha^2 \frac{\mathcal{A}\nabla u_\delta \cdot \nabla u_\delta}{u_\delta^2} dm \\ &\leq C \int_{\mathbb{R}^n} \alpha\,(1-u_\delta)\,\frac{\mathcal{A}\nabla u_\delta \cdot \nabla\alpha}{u_\delta}\, dm \\ &\leq C \int_{\mathbb{R}^n} \alpha\,(1-u_\delta)\,\frac{|\nabla u_\delta||\nabla\alpha|}{u_\delta}\, dm \\ &\leq C \left(\int_{\mathbb{R}^n} \alpha^2 \frac{|\nabla u_\delta|^2}{u_\delta^2}\, dm\right)^{\frac{1}{2}} \left(\int_{\mathbb{R}^n} (1-u_\delta)^2\, |\nabla\alpha|^2\, dm\right)^{\frac{1}{2}} \end{aligned} \tag{8.80}$$

by Cauchy-Schwarz' inequality. Therefore,

$$\int_{\mathbb{R}^n} \alpha^2 |\nabla \ln u_\delta|^2 dm = \int_{\mathbb{R}^n} \alpha^2 \frac{|\nabla u_\delta|^2}{u_\delta^2} dm \leq C \int_{\mathbb{R}^n} (1-u_\delta)^2 |\nabla\alpha|^2 dm \leq C \int_{\mathbb{R}^n} |\nabla\alpha|^2 dm \tag{8.81}$$

because $0 \leq u_\delta \leq 1$, and then with our particular choice of α,

$$m^{-1}(2B) \int_{2B} |\nabla v_\delta|^2 dm = m^{-1}(2B) \int_{2B} |\nabla \ln u_\delta|^2 dm \leq \frac{C}{r^2}. \tag{8.82}$$

We inject this last estimate in (8.78) and get that

$$\sup_B v_\delta = \sup_B (-\ln u_\delta) \leq C, \tag{8.83}$$

i.e. $\inf_B u_\delta = \inf_B \min\{1, u+\delta\} \geq e^{-C} = C^{-1}$. Since the constant doesn't depend on δ, we have the right conclusion, that is $\inf_B u \geq C^{-1}$. □

LEMMA 8.15 (Oscillation estimates on the boundary). *Let B be a ball centered on Γ and $u \in W_r(4B)$ be a solution in $4B \setminus \Gamma$ such that Tu is uniformly bounded on $4B$. Then, there exists $\eta \in (0,1)$ such that*

$$\operatorname*{osc}_B u \leq \eta \operatorname*{osc}_{4B} u + (1-\eta) \operatorname*{osc}_{\Gamma\cap 4B} Tu. \tag{8.84}$$

The constant η depends only on the dimensions n, d and the constants C_0 and C_1.

PROOF. Set $M_4 = \sup_{4B} u$, $m_4 = \inf_{4B} u$, $M_1 = \sup_B u$, $m_1 = \inf_B u$, $M = \sup_{4B\cap\Gamma} Tu$ and $m = \inf_{4B\cap\Gamma} Tu$. Let us first prove that

$$M_4 - M_1 \geq c(M_4 - M) \tag{8.85}$$

and

$$m_1 - m_4 \geq c(m - m_4) \tag{8.86}$$

for some $c \in (0,1]$. Notice that (8.85) is trivially true if $M_4 - M = 0$. Otherwise, we apply Lemma 8.14 to the non-negative supersolution $\min\{\frac{M_4-u}{M_4-M}, 1\}$ whose trace equals 1 on $4B$ (with Lemma 6.1) and we obtain for some constant $c \in (0,1]$

$$\frac{M_4 - M_1}{M_4 - M} \geq c \tag{8.87}$$

which gives (8.85) if we multiply both sides by $M_4 - M$. In the same way, (8.86) is true if $m - m_4 = 0$ and otherwise, we apply Lemma 8.14 to the non-negative supersolution $\min\{\frac{u-m_4}{m-m_4}, 1\}$ and we get for some $c \in (0,1]$

$$\frac{m_1 - m_4}{m - m_4} \geq c, \tag{8.88}$$

which is (8.86).

We sum then (8.85) and (8.86) to get

$$[M_4 - m_4] - [M_1 - m_1] \geq c[M_4 - m_4] - c[M - m], \tag{8.89}$$

that is

$$[M_1 - m_1] \leq (1-c)[M_4 - m_4] + c[M - m], \tag{8.90}$$

which is exactly the desired result. □

We end the section with the Hölder continuity of solutions at the boundary.

LEMMA 8.16. *Let $B = B(x,r)$ be a ball centered on Γ and $u \in W_r(B)$ be a solution in B such that Tu is continuous and bounded on B. There exists $\alpha > 0$ such that for $0 < s < r$,*

$$\operatorname*{osc}_{B(x,s)} u \leq C\left(\frac{s}{r}\right)^{\alpha} \operatorname*{osc}_{B(x,r)} u + C \operatorname*{osc}_{B(x,\sqrt{sr})\cap\Gamma} Tu \tag{8.91}$$

where the constants α, C depend only on the dimensions n and d and the constants C_0 and C_1. In particular, u is continuous on B.

If, in addition, $Tu \equiv 0$ on B, then for any $0 < s < r/2$

$$\operatorname*{osc}_{B(x,s)} u \leq C\left(\frac{s}{r}\right)^{\alpha} \left(m(B)^{-1} \int_B |u|^2 dm\right)^{\frac{1}{2}}. \tag{8.92}$$

PROOF. The first part of the Lemma, i.e. the estimate (8.91), is a straightforward consequence of Lemma 8.15 and [**GT**, Lemma 8.23]. Basically, [**GT**, Lemma 8.23] is a result on functions stating that the functional inequality (8.84) can be turned, via iterations, into (8.91).

The second part of the Lemma is simply a consequence of the first part and of the Moser inequality given in Lemma 8.12. □

CHAPTER 9

Harmonic Measure

We want to solve the Dirichlet problem

$$\left\{\begin{array}{ll} Lu = f & \text{in } \Omega \\ u = g & \text{on } \Gamma, \end{array}\right. \tag{9.1}$$

with a notation that we explain now. Here we require u to lie in W, and by the second line we actually mean that $Tu = g$ σ-almost everywhere on Γ, where T is our trace operator. Logically, we are only interested in functions $g \in H$, because we know that $T(u) \in H$ for $u \in W$.

The condition $Lu = f$ in Ω is taken in the weak sense, i.e. we say that $u \in W$ satisfies $Lu = f$, where $f \in W^{-1} = (W_0)^*$, if for any $v \in W_0$,

$$a(u,v) = \int_\Omega A\nabla u \cdot \nabla v = \langle f, v\rangle_{W^{-1},W_0}. \tag{9.2}$$

Notice that when $f \equiv 0$, a function $u \in W$ that satisfies (9.2) is a solution in Ω.

Now, we made sense of (9.1) for at least $f \in W^{-1}$ and $g \in H$. The next result gives a good solution to the Dirichlet problem.

LEMMA 9.1. *For any $f \in W^{-1}$ and any $g \in H$, there exists a unique $u \in W$ such that*

$$\left\{\begin{array}{ll} Lu = f & \textit{in } \Omega \\ Tu = g & \textit{a.e. on } \Gamma. \end{array}\right. \tag{9.3}$$

Moreover, there exists $C > 0$ independent of f and g such that

$$\|u\|_W \leq C(\|g\|_H + \|f\|_{W^{-1}}), \tag{9.4}$$

where

$$\|f\|_{W^{-1}} = \sup_{\substack{\varphi \in W_0 \\ \|\varphi\|_W = 1}} \langle f, \varphi\rangle_{W^{-1},W_0}. \tag{9.5}$$

PROOF. Since $g \in H$, Theorem 7.3 implies that there exists $G \in W$ such that $T(G) = g$ and

$$\|G\|_W \leq C\|g\|_H. \tag{9.6}$$

The quantity LG is an element of W^{-1} defined by

$$\langle LG, \varphi\rangle_{W^{-1},W_0} := \int_\Omega A\nabla G \cdot \nabla\varphi \, dz = \int_\Omega \mathcal{A}\nabla G \cdot \nabla \varphi \, dm, \tag{9.7}$$

and notice that

$$\|LG\|_{W^{-1}} \leq C\|G\|_W \leq C\|g\|_H \tag{9.8}$$

by (8.8) and (9.6).

Observe that the conditions (8.8) and (8.9) imply that the bilinear form a is bounded and coercive on W_0. It follows from the Lax-Milgram theorem that there exists a (unique) $v \in W_0$ such that $Lv = -LG - f$. Set $u = G - v$. It is now easy to see that $Tu = g$ a.e. on Γ and $Lu = f$ in Γ. The existence of a solution of (9.3) follows.

It remains to check the uniqueness of the solution and the bounds (9.4). Take $u_1, u_2 \in W$ two solutions of (9.3). One has then $T(u_1 - u_2) = g - g = 0$ and hence $u_1 - u_2 \in W_0$. Moreover, $L(u_1 - u_2) = 0$. Since a is bounded and coercive, the uniqueness in the Lax-Milgram theorem yields $u_1 - u_2 = 0$. Therefore (9.3) has also a unique solution.

Finally, let us prove the bounds (9.4). From the coercivity of a, we get that

$$\|v\|_W^2 \leq Ca(v,v) \leq C\|LG + f\|_{W^{-1}}\|v\|_W, \tag{9.9}$$

i.e., with (9.8),

$$\|v\|_W \leq C\|LG + f\|_{W^{-1}} \leq C(\|g\|_H + \|f\|_{W^{-1}}). \tag{9.10}$$

We conclude the proof of (9.4) with

$$\|u\|_W = \|G - v\|_W \leq C(\|g\|_H + \|f\|_{W^{-1}}) \tag{9.11}$$

by (9.6). □

The next step in the construction of a harmonic measure associated to L, is to prove a maximum principle.

LEMMA 9.2. *Let $u \in W$ be a supersolution in Ω satisfying $Tu \geq 0$ a.e. on Γ. Then $u \geq 0$ a.e. in Ω.*

PROOF. Set $v = \min\{u, 0\} \leq 0$. According to Lemma 6.1 (b), we have

$$\nabla v = \begin{cases} \nabla u & \text{if } u < 0 \\ 0 & \text{if } u \geq 0 \end{cases} \tag{9.12}$$

and

$$Tv = \min\{Tu, 0\} = 0 \quad \text{a.e. in } \Gamma. \tag{9.13}$$

In particular, (9.13) implies that $v \in W_0$. The third case of Lemma 8.3 allows us to test v against the supersolution $u \in W$; this gives

$$\int_\Omega \mathcal{A}\nabla u \cdot \nabla v \, dm \leq 0, \tag{9.14}$$

that is with (9.12),

$$\int_\Omega \mathcal{A}\nabla v \cdot \nabla v \, dm = \int_{\{u<0\}} \mathcal{A}\nabla u \cdot \nabla u \, dm = \int_\Omega \mathcal{A}\nabla u \cdot \nabla v \, dm \leq 0. \tag{9.15}$$

Together with the ellipticity condition (8.9), we obtain $\|v\|_W \leq 0$. Recall from Lemma 5.2 that $\|.\|_W$ is a norm on $W_0 \ni v$, hence $v = 0$ a.e. in Ω. We conclude from the definition of v that $u \geq 0$ a.e. in Ω. □

Here is a corollary of Lemma 9.2.

LEMMA 9.3 (Maximum principle). *Let $u \in W$ be a solution of $Lu = 0$ in Ω. Then*

$$\sup_\Omega u \leq \sup_\Gamma Tu \tag{9.16}$$

and

$$\inf_{\Omega} u \geq \inf_{\Gamma} Tu, \tag{9.17}$$

where we recall that sup *and* inf *actually essential supremum and infimum. In particular, if* Tu *is essentially bounded,*

$$\sup_{\Omega} |u| \leq \sup_{\Gamma} |Tu|. \tag{9.18}$$

PROOF. Let us prove (9.16). Write M for the essential supremum of Tu on Γ; we may assume that $M < +\infty$, because otherwise (9.16) is trivial. Then $M - u \in W$ and $T(M - u) \geq 0$ a.e. on Γ. Lemma 9.2 yields $M - u \geq 0$ a.e. in Ω, that is

$$\sup_{\Omega} u \leq \sup_{\Gamma} Tu. \tag{9.19}$$

The lower bound (9.17) is similar, and (9.18) follows. □

We want to define the harmonic measure via the Riesz representation theorem (for measures), that requires a linear form on the space of compactly supported continuous functions on Γ. We denote this space by $C_0^0(\Gamma)$; that is, $g \in C_0^0(\Gamma)$ if g is defined and continuous on Γ, and there exists a ball $B \subset \mathbb{R}^n$ centered on Γ such that $\operatorname{supp} g \subset B \cap \Gamma$.

LEMMA 9.4. *There exists a bounded linear operator*

$$U : C_0^0(\Gamma) \to C^0(\mathbb{R}^n) \tag{9.20}$$

such that, for every every $g \in C_0^0(\Gamma)$,

(i) *the restriction of* Ug *to* Γ *is* g;
(ii) $\sup_{\mathbb{R}^n} Ug = \sup_{\Gamma} g$ *and* $\inf_{\mathbb{R}^n} Ug = \inf_{\Gamma} g$;
(iii) $Ug \in W_r(\Omega)$ *and is a solution of* L *in* Ω;
(iv) *if* B *is a ball centered on* Γ *and* $g \equiv 0$ *on* B, *then* Ug *lies in* $W_r(B)$;
(v) *if* $g \in C_0^0(\Gamma) \cap H$, *then* $Ug \in W$, *and it is the solution of* (9.3), *with* $f = 0$, *provided by Lemma* 9.1.

PROOF. This is essentially an argument of extension from a dense class by uniform continuity. We first define U on $C_0^0(\Gamma) \cap H$, by saying that $u = Ug$ is the solution of (9.3), with $f = 0$, provided by Lemma 9.1. Thus $u \in W$; but since its trace is $Tu = g$ is continuous, it follows from Lemmata 8.8 and 8.16 (the Hölder continuity inside and at the boundary) that u is continuous on $\mathbb{R}^n$.

Next we check that U is linear and bounded on $C_0^0(\Gamma) \cap H \subset C_0^0(\Gamma)$ (where we use the sup norm). The linearity comes from the uniqueness in Lemma 9.1, and the boundedness from the maximum principle: for $g, h \in C_0^0(\Gamma) \cap H$, we can apply (9.19) to $u = Ug - Uh$, and we get that $\sup_{\mathbb{R}^n} |u| = \sup_{\Omega} |u| \leq \sup_{\Gamma} |Tu| = ||g - h||_{\infty}$ because u is continuous.

It is clear that $C_0^0(\Gamma) \cap H$ is dense in $C_0^0(\Gamma)$, because (restrictions to Γ of) compactly supported smooth functions on $\mathbb{R}^n$ (or even Lipschitz functions, for that matter) lie in H: compute their norm in (1.5) directly. Thus U has a unique extension by continuity to $C_0^0(\Gamma)$. We could even define U, with the same properties, on its closure (continuous functions that tend to 0 at infinity), but we decided not to bother.

We are now ready to check the various properties of U. Given $g \in C_0^0(\Gamma)$, select a sequence (g_k) of compactly supported smooth functions that converges to g in the sup norm. Then $u_k = Ug_k$ converges uniformly in $\mathbb{R}^n$ to $u = Ug$, and in particular u is continuous and its restriction to Γ is g, as in (i). In addition, (ii) holds because $\sup_{\mathbb{R}^n} u = \lim_{k\to+\infty} \sup_{\mathbb{R}^n} u_k \leq \lim_{k\to+\infty} \sup_\Gamma g_k = \sup_\Gamma g$, and similarly for the infimum.

For (iii) we first need to check that $u \in W_r(\Omega)$. Observe that we know these facts for the u_k, so we'll only need to take limits. Let $\phi \in C_0^\infty(\Omega)$ be given. Lemma 8.6 (Caccioppoli's inequality) says that, since u_k is a solution,

(9.21)

$$\int_\Omega |\nabla(\phi u_k)|^2 dm \leq C\int_\Omega \phi^2|\nabla u_k|^2 dm + C\int_\Omega |\nabla\phi|^2|u_k|^2 dm \leq C\int_\Omega |\nabla\phi|^2|u_k|^2 dm.$$

The right-hand side of (9.21) converges to $C\int_\Omega |\nabla\phi|^2|u|^2 dm$, since $|\nabla\phi|^2$ is bounded and compactly supported. So $\int_B |\nabla(\phi u_k)|^2 dm$ is bounded uniformly in k. Since the ϕu_k vanish outside of the support of ϕ (which lies far from Γ) and converge uniformly to ϕu, we get that the ϕu_k converge to ϕu in L^1 and, since the $|\nabla(\phi u_k)|$ are uniformly bounded in $L^2(\Omega, w)$, we can find a subsequence for which they converge weakly to a limit $V \in L^2(\Omega, w)$. We easily check on test functions that $\nabla(\phi u) = V$, hence $\phi u \in W$ for any $\phi \in C_0^\infty(\Omega)$, and $u \in W_r(\Omega)$.

Next we check that u is a solution in Ω, i.e., that for $\varphi \in C_0^\infty(\Omega)$,

$$\int_\Omega \mathcal{A}\nabla u \cdot \nabla\varphi\, dm = 0. \tag{9.22}$$

Let $\varphi \in C_0^\infty(\Omega)$ be given, and choose $\phi \in C_0^\infty(\Omega)$ such that $\phi \equiv 1$ on $\operatorname{supp}\varphi$. We just proved that for some subsequence, $\nabla(\phi u_k)$ converges weakly to $\nabla(\phi u)$ in $L^2(\Omega, w)$. Since u_k is a solution for every k,

$$\begin{aligned}\int_\Omega \mathcal{A}\nabla u \cdot \nabla\varphi\, dm &= \int_\Omega \mathcal{A}\nabla(\phi u) \cdot \nabla\varphi\, dm = \lim_{k\to\infty}\int_\Omega \mathcal{A}\nabla(\phi u_k) \cdot \nabla\varphi\, dm \\ &= \lim_{k\to\infty}\int_\Omega \mathcal{A}\nabla u_k \cdot \nabla\varphi\, dm = 0. \end{aligned} \tag{9.23}$$

This proves (9.22) and (iii) follows.

For (iv), suppose in addition that $g \equiv 0$ on a ball B centered on Γ; we want similar results in B (that is, across Γ). Notice that it is easy to approximate it (in the supremum norm) by smooth, compactly supported functions g_k that also vanish on $\Gamma \cap B$. Let use such a sequence (g_k) to define $Ug = \lim_{k\to+\infty} Ug_k$.

Let $\varphi \in C_0^\infty(B)$ be given, and let us check that $\varphi u \in W$. Set $K = \operatorname{supp}\varphi$, suppose $K \neq \emptyset$, and set $\delta = \operatorname{dist}(K, \partial B) > 0$. Cover $K \cap \Gamma$ by a finite number of balls balls B_i of radius $10^{-1}\delta$ centered on $K \cap \Gamma$, and then cover $K' = K \setminus \cup_i B_i$ by a finite number of balls B_j of radius $10^{-2}\delta$ centered on that set K'. We can use a partition of unity composed of smooth functions supported in the $2B_i$ and the $2B_j$ to reduce to the case when φ is supported on a $2B_i$ or a $2B_j$.

Suppose for instance that φ is supported in $2B_i$. We can apply Lemma 8.11 (Caccioppoli's inequality at the boundary) to $u_k = Ug_k$ on the ball $2B_i$, because its trace g_k vanishes on $4B_i$. We get that

$$\int_{2B_i} |\nabla(\varphi u_k)|^2 dm \leq C\int_{2B_i} (|\varphi\nabla u_k)|^2 + |u_k\nabla\varphi|^2) dm \leq \int_{4B_i} |\nabla\varphi|^2|u_k|^2 dm. \tag{9.24}$$

With this estimate, we can proceed as with (9.21) above to prove that $\varphi u \in W$ and its derivative is the weak limit of the $\nabla(\varphi u_k)$. When instead φ is supported in a $2B_j$, we use the interior Caccioppoli inequality (Lemma 8.6 and proceed as above).

Thus $u = Ug$ lies in $W_r(B)$, and this proves (iv). We started the proof with (v), so this completes our proof of Lemma 9.4. □

Our next step is the construction of the harmonic measure. Let $X \in \Omega$. By Lemma 9.4, the linear form

$$g \in C_0^0(\Gamma) \to Ug(X) \tag{9.25}$$

is bounded and positive (because $u = Ug$ is nonnegative when $g \geq 0$). The following statement is thus a direct consequence of the Riesz representation theorem (see for instance [**Rud**, Theorem 2.14]).

LEMMA 9.5. *There exists a unique positive regular Borel measure ω^X on Γ such that*

$$Ug(X) = \int_\Gamma g(y) d\omega^X(y) \tag{9.26}$$

for any $g \in C_0^0(\Gamma)$. Besides, for any Borel set $E \subset \Gamma$,

$$\omega^X(E) = \sup\{\omega^X(K) : E \supset K,\ K \text{ compact}\} = \inf\{\omega^X(V) : E \subset V,\ V \text{ open}\}. \tag{9.27}$$

The harmonic measure is a probability measure, as proven in the following result.

LEMMA 9.6. *For any $X \in \Omega$,*

$$\omega^X(\Gamma) = 1.$$

PROOF. Let $X \in \Omega$ be given. Choose $x \in \Gamma$ such that $\delta(X) = |X - x|$. Set then $B_j = B(x, 2^j \delta(X))$. According to (9.27),

$$\omega^X(\Gamma) = \lim_{j \to +\infty} \omega^X(\overline{B_j}). \tag{9.28}$$

Choose, for $j \geq 1$, $\bar{g}_j \in C_0^\infty(B_{j+1})$ such that $0 \leq \bar{g}_j \leq 1$ and $\bar{g}_j \equiv 1$ on $\overline{B_j}$ and then define $g_j = T(\bar{g}_j)$. Since the harmonic measure is positive, we have

$$\omega^X(B_j) \leq \int_\Gamma g_j(y) d\omega^X(y) \leq \omega^X(B_{j+1}). \tag{9.29}$$

Together with (9.28),

$$\omega^X(\Gamma) = \lim_{j \to +\infty} \int_\Gamma g_j(y) d\omega^X(y) = \lim_{j \to +\infty} u_j(X), \tag{9.30}$$

where u_j is the image by the map (9.20) of the function g_j. Since g_j is the trace of a smooth and compactly supported function, $g_j \in H$ and so $u_j \in W$ is the solution of (9.3) with data g_j. Moreover, $0 \leq u_j \leq 1$ by Lemma 9.4 (ii). We want to show that $u_j(X) \to 1$ when $j \to +\infty$. The function $v_j := 1 - u_j \in W$ is a solution in B_j satisfying $Tv_j \equiv 0$ on B_j. So Lemma 8.16 says that

$$0 \leq 1 - u_j(X) = v_j(X) \leq \operatorname*{osc}_{B_1} v_j \leq C 2^{-j\alpha} \operatorname*{osc}_{B_j} v_j \leq C 2^{-j\alpha}, \tag{9.31}$$

where $C > 0$ and $\alpha > 0$ are independent of j. It follows that $v_j(X)$ tends to 0, and $u_j(X)$ tends to 1 when j goes to $+\infty$. The lemma follows from this and (9.30), the lemma follows. □

LEMMA 9.7. *Let $E \subset \Gamma$ be a Borel set and define the function u_E on Ω by $u_E(X) = \omega^X(E)$. Then*

(i) *if there exists* $X \in \Omega$ *such that* $u_E(X) = 0$, *then* $u_E \equiv 0$;
(ii) *the function* u_E *lies in* $W_r(\Omega)$ *and is a solution in* Ω;
(iii) *if* $B \subset \mathbb{R}^n$ *is a ball such that* $E \cap B = \emptyset$, *then* $u_E \in W_r(B)$ *and* $Tu_E = 0$ *on* B.

PROOF. First of all, $0 \leq u_E \leq 1$ because ω^X is a positive probability measure for any $X \in \Omega$.

Let us prove (i). Thanks to (9.27), it suffices to prove the result when $E = K$ is compact.

Let $X \in \Omega$ be such that $u_K(X) = 0$. Let $Y \in \Omega$ and $\epsilon > 0$ be given. By (9.27) again, we can find an open U such that $U \supset K$ and $\omega^X(U) < \epsilon$. Urysohn's lemma (see for instance Lemma 2.12 in [**Rud**]) gives the existence of $g \in C_0^0(\Gamma)$ such that $0 \leq g \leq 1$ and $g \equiv 1$ on K. Set $u = Ug$, where U is as in (9.20). Thanks to the positivity of the harmonic measure, $u_K \leq u$. Let $Y \in \Omega$ be given, and apply the Harnack inequality (8.35) to u (notice that u lies in $W_r(\Omega)$ and is a solution in Ω thanks to Lemma 9.4). We get that

$$0 \leq u_K(Y) \leq u(Y) \leq C_{X,Y} u(X) \leq C_{X,Y}\epsilon. \tag{9.32}$$

Since (9.32) holds for any positive ϵ, we have $u_K(Y) = 0$. Part (i) of the lemma follows.

We turn to the proof of (ii), which we first do when $E = V$ is open. We first check that

$$u_V \text{ is a continuous function on } \Omega. \tag{9.33}$$

Fix $X \in \Omega$, and build an increasing sequence of compact sets $K_j \subset V$ such that $\omega^X(V) < \omega^X(K_j) + \frac{1}{j}$. With Urysohn's lemma again, we construct $g_j \in C_0^0(V)$ such that $\mathbb{1}_{K_j} \leq g_j \leq \mathbb{1}_V$ and, without loss of generality we can choose $g_j \leq g_i$ whenever $j \leq i$. Set $u_j = Ug_j \in C^0(\mathbb{R}^n)$, as in (9.20), and notice that $u_j(X) = \int_\Gamma g_j d\omega^X$ by (9.26). Then for $j \geq 1$,

$$u_{K_j}(X) = \omega^X(K_j) \leq u_j(X) \leq \omega^X(V) = u_V(X) \leq \omega^X(K_j) + \frac{1}{j} \tag{9.34}$$

by definition of u_E, because the harmonic measure is nondecreasing, and since $\mathbb{1}_{K_j} \leq g_j \leq \mathbb{1}_V$. Similarly, (u_j) is a nondecreasing sequence of functions, i.e.,

$$u_i \geq u_j \text{ on } \Omega \text{ for } i \geq j \geq 1, \tag{9.35}$$

by the maximum principle in Lemma 9.4 and because $g_i \geq g_j$, so that in particular

$$u_j(X) \leq u_i(X) \leq u_j(X) + \frac{1}{j} \quad \text{for } i \geq j \geq 1, \tag{9.36}$$

by (9.34). Now $u_i - u_j$ is a nonnegative solution (by Lemma 9.4), and Lemma 8.10 implies that for every compact set $J \subset \Omega$, there exists $C_J > 0$ such that

$$0 \leq \sup_J (u_i - u_j) \leq C_J(u_i - u_j)(X) \leq \frac{C_J}{j} \tag{9.37}$$

for $i \geq j \geq 1$. We deduce from this that $(u_j)_j$ converges uniformly on compact sets of Ω to a function u_∞, which is therefore continuous on Ω. Thus (9.33) will follow as soon as we prove that $u_\infty = u_V$.

Set $K = \bigcup_j K_j$; then $u_{K_j} \leq u_K \leq u_V$ by monotonicity of the harmonic measure, and (9.34) implies that $u_K(X) = u_V(X)$. Now $u_V - u_K = u_{V\setminus K}$, so $u_{V\setminus K}(X) = 0$. By Point (i) of the present lemma, $u_{V\setminus K}(Y) = 0$ for every $Y \in \Omega$.

But $u_V(Y) = \omega^Y(V)$, and ω^Y is a measure, so $u_{V\setminus K}(Y) = \lim_{j\to+\infty} u_{V\setminus K_j}(Y) = u_V(Y) - \lim_{j\to+\infty} u_{K_j}(Y)$.

Since $u_{K_j}(Y) \le u_j(Y) \le u_V(Y)$ by the proof of (9.34), we get that $u_j(Y)$ tends to $u_V(Y)$. In other words, $u_\infty(Y) = u_V(Y)$, and (9.33) follows as announced.

We proved that u_V is continuous on Ω and that it is the limit, uniformly on compact subsets of Ω, of a sequence of functions $u_j \in C^0(\mathbb{R}^n) \cap W_r(\Omega)$, which are also solutions of L in Ω. We now want to prove that $u_V \in W_r(\Omega)$, and we proceed as we did near (9.21).

Let $\phi \in C_0^\infty(\Omega)$ be given. In the distributional sense, we have $\nabla(\phi u_j) = u_j \nabla\phi + \phi\nabla u_j$. So the Caccioppoli inequality given by Lemma 8.6 yields

$$(9.38)\quad \int_\Omega |\nabla(\phi u_j)|^2 dm \le C\int_\Omega (|\nabla\phi|^2|u_j|^2 + \phi^2|\nabla u_j|^2) dm \le C\int_\Omega |\nabla\phi|^2|u_j|^2 dm.$$

Since the u_j converge to u uniformly on $\operatorname{supp}\phi$, the right-hand side of (9.38) converges to $C\int_\Omega |\nabla\phi|^2|u|^2 dm$. Consequently, the left-hand side of (9.38) is uniformly bounded in j and hence there exists $v \in L^2(\Omega, w)$ such that $\nabla(\phi u_j)$ converges weakly to v in $L^2(\Omega, w)$. By uniqueness of the limit, the distributional derivative $\nabla(\phi u_V)$ equals $v \in L^2(\Omega, w)$, so by definition of W, $\phi u_V \in W$. Since the result holds for any $\phi \in C_0^\infty(\Omega)$, we just established $u_V \in W_r(\Omega)$ as desired. In addition, we also checked that (for a subsequence) $\nabla(\phi u_j)$ converges weakly in $L^2(\Omega, w)$ to $\nabla(\phi u_V)$.

We now establish that u_V is a solution. Let $\varphi \in C_0^\infty(\Omega)$ be given. Choose $\phi \in C_0^\infty(\Omega)$ such that $\phi \equiv 1$ on $\operatorname{supp}\varphi$. Thanks to the weak convergence of $\nabla(\phi u_j)$ to $\nabla(\phi u_j)$

$$(9.39)\quad \begin{aligned} \int_\Omega \mathcal{A}\nabla u_V \cdot \nabla\varphi\, dm &= \int_\Omega \mathcal{A}\nabla(\phi u_V)\cdot\nabla\varphi\, dm \\ &= \lim_{j\to+\infty}\int_\Omega \mathcal{A}\nabla(\phi u_j)\cdot\nabla\varphi\, dm = \lim_{j\to+\infty}\int_\Omega \mathcal{A}\nabla u_j\cdot\nabla\varphi\, dm = 0 \end{aligned}$$

because each u_j is a solution. Hence u_V is a solution.

This completes our proof of (ii) when $E = V$ is open. The proof of (ii) for general Borel sets E works similarly, but we now approximate E from above by open sets. Fix $X \in \Omega$. Thanks to the regularity property (9.27), there exists a decreasing sequence (V_j) of open sets that contain E, and for which $u_{V_j}(X)$ tends to $u_E(X)$.

From our previous work, we know that each u_{V_j} is continuous on Ω, lies in $W_r(\Omega)$, and is a solution in Ω. Using the same process as before, we can show first that the u_{V_j} converge, uniformly on compact sets of Ω, to u_E, which is then continuous on Ω. Then we prove that, for any $\phi \in C_0^\infty(\Omega)$, $\nabla(\phi u_{V_j})$ converges weakly in $L^2(\Omega, w)$ to $\nabla(\phi u_E)$, from which we deduce $u_E \in W_r(\Omega)$ and then that u_E is a solution.

Part (iii) of the lemma remains to be proven. Let $B \subset \mathbb{R}^n$ be a ball such that $B \cap E = \emptyset$. Since u_E lies in $W_r(\Omega)$ and is a solution, Lemma 8.8 says that u_E is continuous in Ω. We first prove that if we set $u = 0$ on $B\cap\Gamma$, we get a continuous extension of u, (with then has a vanishing trace, or restriction, on $B\cap\Gamma$).

Let $x \in B\cap\Gamma$ be given. Choose $r > 0$ such that $B(x, 2r) \subset B$ and then construct a function $\bar g \in C_0^\infty(B(x, 2r))$ such that $\bar g \equiv 1$ in $B(x, r)$. Since $\bar g$ is smooth and compactly supported, $g := T(\bar g)$ lies in $H \cap C_0^0(\Gamma)$ and then $u = Ug$, the image of g by the map of (9.20), lies in in $W \cap C^0(\mathbb{R}^n)$. From the positivity of the harmonic measure, we deduce that $0 \leq u_E \leq 1 - u$. Since 0 and $1 - u$ are both continuous functions that are equal 0 at x, the squeeze theorem says that u_E is continuous (or can be extended by continuity) at x, and $u_E(x) = 0$.

To complete the proof of the lemma, we show that u_E actually lies in $W_r(B)$. As for the proof of (ii), we first assume that $E = V$ is open. We take a nondecreasing sequence of compact sets $K_j \subset V$ that converges to V, and then we build $g_j \in C_0^0(V)$, such that $\mathbb{1}_{K_j} \leq g_j \leq \mathbb{1}_V$ and the sequence (g_j) is non-decreasing. We then take $u_j = Ug_j$ (with the map from (9.20)), and in particular the sequence (u_j) is non-decreasing on Ω. From the proof of (ii), we know that u_j converges to u_V on compact sets of Ω, then in particular u_j converges pointwise to u_V in Ω.

Let $\varphi \in C_0^\infty(B)$; we want to prove that $\varphi u_V \in W$. From Lemma 8.11, we have

$$\int_B |\nabla(\varphi u_j)|^2 dm \leq C \int_B (|\nabla\varphi|^2 |u_j|^2 + \varphi^2 |\nabla u_j|^2) dm \leq C \int_B |\nabla\varphi|^2 |u_j|^2 dm. \tag{9.40}$$

Since u is continuous on B, $u_V \in L^2(\operatorname{supp}\varphi, w)$ and the right-hand side converges to $C \int_B |\nabla\varphi|^2 |u_V|^2 dm$ by the dominated convergence theorem. The left-hand side is thus uniformly bounded in j and $\nabla(\varphi u_j)$ converges weakly, maybe after extracting a subsequence, to some v in $L^2(B, w)$. By uniqueness of the limit, $v = \nabla(\varphi u_V) \in L^2(B, w)$. Since the result holds for all $\varphi \in C_0^\infty(B)$, we get $u_V \in W_r(B)$.

In the general case where E is a Borel set, fix $X \in \Omega$ and take a decreasing sequence of open sets $V_j \supset X$ such that $u_{V_j}(X) \to u_E(X)$. We can prove using part (i) of this lemma that u_{V_j} converges to u_E pointwise in Ω. Then we use Lemma 8.11 to show that for $\varphi \in C_0^\infty(B)$,

$$\int_B |\nabla(\varphi u_{V_j})|^2 dm \leq C \int_B |\nabla\varphi|^2 |u_{V_j}|^2 dm \tag{9.41}$$

when j is so large that V_j is far from the support of φ. The right-hand side has a limit, thanks to the dominated convergence theorem, thus the left-hand side is uniformly bounded in j. So there exists a subsequence of $\nabla(\varphi u_{V_j})$ that converges weakly in $L^2(B, w)$, and by uniqueness to the limit, the limit has to be $\nabla(\varphi u_E)$, which thus lies in $L^2(B, w)$. We deduce that $\varphi u_E \in W$ and then $u \in W_r(B)$. $\square$

CHAPTER 10

Green Functions

The aim of this chapter is to define a Green function, that is, formally, a function g defined on $\Omega \times \Omega$ and such that for $y \in \Omega$,

$$\begin{cases} Lg(.,y) = \delta_y & \text{in } \Omega \\ Tg(.,y) = 0 & \text{on } \Gamma. \end{cases} \tag{10.1}$$

where δ_y denotes the Dirac distribution.

Our proof of existence and uniqueness, and the estimates below, are adapted from arguments of [**GW**] (see also [**HoK**] and [**DK**]) for the classical case of codimension 1.

LEMMA 10.1. *There exists a non-negative function $g: \Omega \times \Omega \to \mathbb{R} \cup \{+\infty\}$ with the following properties.*

(i) *For any $y \in \Omega$ and any function $\alpha \in C_0^\infty(\mathbb{R}^n)$ such that $\alpha \equiv 1$ in a neighborhood of y*

$$(1-\alpha)g(.,y) \in W_0. \tag{10.2}$$

In particular, $g(.,y) \in W_r(\mathbb{R}^n \setminus \{y\})$ and $T[g(.,y)] = 0$.

(ii) *For every choice of $y \in \Omega$, $R > 0$, and $q \in [1, \frac{n}{n-1})$,*

$$g(.,y) \in W^{1,q}(B(y,R)) := \{u \in L^q(B(y,R)),\ \nabla u \in L^q(B(y,R))\}. \tag{10.3}$$

(iii) *For $y \in \Omega$ and $\varphi \in C_0^\infty(\Omega)$,*

$$\int_\Omega A\nabla_x g(x,y) \cdot \nabla\varphi(x)\,dx = \varphi(y). \tag{10.4}$$

In particular, $g(.,y)$ is a solution of $Lu = 0$ in $\Omega \setminus \{y\}$.

In addition, the following bounds hold.

(iv) *For $r > 0$, $y \in \Omega$ and $\epsilon > 0$,*

$$\int_{\Omega\setminus B(y,r)} |\nabla_x g(x,y)|^2 dm(x) \leq \begin{cases} Cr^{1-d} & \text{if } 4r \geq \delta(y) \\ \frac{Cr^{2-n}}{w(y)} & \text{if } 2r \leq \delta(y),\ n \geq 3 \\ \frac{C_\epsilon}{w(y)}\left(\frac{\delta(y)}{r}\right)^\epsilon & \text{if } 2r \leq \delta(y),\ n = 2, \end{cases} \tag{10.5}$$

where $C > 0$ depends on d, n, C_0, C_1 and $C_\epsilon > 0$ depends on d, C_0, C_1, and ϵ.

(v) *For $x, y \in \Omega$ such that $x \neq y$ and $\epsilon > 0$,*

$$0 \leq g(x,y) \leq \begin{cases} C|x-y|^{1-d} & \text{if } 4|x-y| \geq \delta(y) \\ \frac{C|x-y|^{2-n}}{w(y)} & \text{if } 2|x-y| \leq \delta(y),\ n \geq 3 \\ \frac{C_\epsilon}{w(y)}\left(\frac{\delta(y)}{|x-y|}\right)^\epsilon & \text{if } 2|x-y| \leq \delta(y),\ n = 2, \end{cases} \tag{10.6}$$

where again $C > 0$ depends on d, n, C_0, C_1 and $C_\epsilon > 0$ depends on d, C_0, C_1, ϵ.

(vi) *For $q \in [1, \frac{n}{n-1})$ and $R \geq \delta(y)$,*

$$\int_{B(y,R)} |\nabla_x g(x,y)|^q dm(x) \leq C_q R^{d(1-q)+1}, \tag{10.7}$$

where $C_q > 0$ depends on d, n, C_0, C_1, and q.

(vii) *For $y \in \Omega$, $R \geq \delta(y)$, $t > 0$ and $p \in [1, \frac{2n}{n-2}]$ (if $n \geq 3$) or $p \in [1, +\infty)$ (if $n = 2$),*

$$\frac{m(\{x \in B(y,R),\, g(x,y) > t\})}{m(B(y,R))} \leq C_p \left(\frac{R^{1-d}}{t}\right)^{\frac{p}{2}}, \tag{10.8}$$

where $C_p > 0$ depends on d, n, C_0, C_1 and p.

(viii) *For $y \in \Omega$, $t > 0$ and $\eta \in (0,2)$,*

$$m(\{x \in \Omega,\, |\nabla_x g(x,y)| > t\}) \leq \begin{cases} Ct^{-\frac{d+1}{d}} & \text{if } t \leq \delta(y)^{-d} \\ Cw(y)^{-\frac{1}{n-1}} t^{-\frac{n}{n-1}} & \text{if } t \geq \delta(y)^{-d},\ n \geq 3 \\ C_\eta w(y)^{-1} \delta(y)^{d\eta} t^{\eta - 2} & \text{if } t \geq \delta(y)^{-d},\ n = 2, \end{cases} \tag{10.9}$$

where $C > 0$ depends on d, n, C_0, C_1 and $C_\eta > 0$ depends on d, C_0, C_1, η.

REMARK 10.2. When $d < 1$ and $|x-y| \geq \frac{1}{2}\delta(y)$, the bound $g(x,y) \leq C|x-y|^{1-d}$ given in (10.6) can be improved into

$$g(x,y) \leq C \min\{\delta(x), \delta(y)\}^{1-d}. \tag{10.10}$$

This fact is proven in Lemma 11.4 below.

REMARK 10.3. The authors believe that the bounds given in (10.5) and (10.6) when $n = 2$ and $2r$ (or $2|x-y|$) is smaller than $\delta(y)$ are not optimal. One should be able to replace for instance the bound $\frac{C_\epsilon}{w(y)}\left(\frac{\delta(y)}{r}\right)^\epsilon$ by $\frac{C}{w(y)} \ln\left(\frac{\delta(y)}{r}\right)$ in (10.5) by adapting the arguments of [**DK**] (see also [**FJK**, Theorem 3.3]). However, the estimates given above are sufficient for our purposes and we didn't want to make this article even longer.

REMARK 10.4. Note that when $n \geq 3$, thanks to Lemma 2.3, the bound (10.6) can be gathered into a single estimate

$$g(x,y) \leq C \frac{|x-y|^2}{m(B(y,|x-y|))} \tag{10.11}$$

whenever $x, y \in \Omega$, $x \neq y$. In the same way, also for $n \geq 3$, the bound (10.5) can be gathered into a single estimate

$$\int_{\Omega \setminus B(y,r)} |\nabla_x g(x,y)|^2 dm(x) \leq C \frac{r^2}{m(B(y,r))} \tag{10.12}$$

whenever $y \in \Omega$ and $r > 0$.

PROOF. This proof will adapt the arguments of [**GW**, Theorem 1.1].

Let $y \in \Omega$ be fixed. Consider again the bilinear form a on $W_0 \times W_0$ defined as

$$a(u,v) = \int_\Omega A\nabla u \cdot \nabla v = \int_\Omega \mathcal{A}\nabla u \cdot \nabla v \, dm. \tag{10.13}$$

The bilinear form a is bounded and coercive on W_0, thanks to (8.8) and (8.9).

Let $\rho > 0$ be small. Take, for instance, ρ such that $100\rho < \delta(y)$. Write B_ρ for $B(y,\rho)$. The linear form

$$\varphi \in W_0 \to \fint_{B_\rho} \varphi(z)\, dz \tag{10.14}$$

is bounded. Indeed, let z be a point in Γ, then

$$\left| \fint_{B_\rho} \varphi(z)\, dz \right| \leq C_{y,z,\rho} \fint_{B(z,|y-z|+\rho)} |\varphi(z)|\, dz \leq C_{y,z,\rho} \|\varphi\|_W \tag{10.15}$$

by Lemma 4.1. By the Lax-Milgram theorem, there exists then a unique function $\mathfrak{g}^\rho = g^\rho(.,y) \in W_0$ such that

$$a(\mathfrak{g}^\rho, \varphi) = \int_\Omega \mathcal{A}\nabla \mathfrak{g}^\rho \cdot \nabla\varphi\, dm = \fint_{B_\rho} \varphi(z)\, dz \qquad \forall \varphi \in W_0. \tag{10.16}$$

We like $\mathfrak{g}^\rho$, and will actually spend some time studying it, because $g(\cdot,y)$ will later be obtained as a limit of the $\mathfrak{g}^\rho$. By (10.16),

$$\mathfrak{g}^\rho \in W_0 \text{ is a solution of } L\mathfrak{g}^\rho = 0 \text{ in } \Omega \setminus \overline{B_\rho}. \tag{10.17}$$

This fact will be useful later on.

For now, let us prove that $\mathfrak{g}^\rho \geq 0$ a.e. on Ω. Since $\mathfrak{g}^\rho \in W_0$, Lemma 6.1 yields $|\mathfrak{g}^\rho| \in W_0$, $\nabla|\mathfrak{g}^\rho| = \nabla \mathfrak{g}^\rho$ a.e. on $\{\mathfrak{g}^\rho > 0\}$, $\nabla|\mathfrak{g}^\rho| = -\nabla\mathfrak{g}^\rho$ a.e. on $\{\mathfrak{g}^\rho < 0\}$ and $\nabla|\mathfrak{g}^\rho| = 0$ a.e. on $\{\mathfrak{g}^\rho = 0\}$. Consequently

(10.18)

$$\begin{aligned} \int_\Omega \mathcal{A}\nabla|\mathfrak{g}^\rho| \cdot \nabla|\mathfrak{g}^\rho|\, dm &= \int_{\{\mathfrak{g}^\rho>0\}} \mathcal{A}\nabla\mathfrak{g}^\rho \cdot \nabla\mathfrak{g}^\rho\, dm + \int_{\{\mathfrak{g}^\rho<0\}} \mathcal{A}\nabla\mathfrak{g}^\rho \cdot \nabla\mathfrak{g}^\rho\, dm \\ &= \int_\Omega \mathcal{A}\nabla\mathfrak{g}^\rho \cdot \nabla\mathfrak{g}^\rho\, dm \end{aligned}$$

and

$$\begin{aligned} \int_\Omega \mathcal{A}\nabla|\mathfrak{g}^\rho| \cdot \nabla\mathfrak{g}^\rho\, dm &= \int_{\{\mathfrak{g}^\rho>0\}} \mathcal{A}\nabla\mathfrak{g}^\rho \cdot \nabla\mathfrak{g}^\rho\, dm - \int_{\{\mathfrak{g}^\rho<0\}} \mathcal{A}\nabla\mathfrak{g}^\rho \cdot \nabla\mathfrak{g}^\rho\, dm \\ &= \int_\Omega \mathcal{A}\nabla\mathfrak{g}^\rho \cdot \nabla|\mathfrak{g}^\rho|\, dm, \end{aligned} \tag{10.19}$$

which can be rewritten $a(|\mathfrak{g}^\rho|, |\mathfrak{g}^\rho|) = a(\mathfrak{g}^\rho, \mathfrak{g}^\rho)$ and $a(|\mathfrak{g}^\rho|, \mathfrak{g}^\rho) = a(\mathfrak{g}^\rho, |\mathfrak{g}^\rho|)$. Moreover, if we use $\mathfrak{g}^\rho \in W_0$ and $|\mathfrak{g}^\rho| \in W_0$ as test functions in (10.16), we obtain

(10.20)

$$a(|\mathfrak{g}^\rho|, |\mathfrak{g}^\rho|) = a(\mathfrak{g}^\rho, \mathfrak{g}^\rho) = \int_{B_\rho} \mathfrak{g}^\rho(z)\, dz \leq \int_{B_\rho} |\mathfrak{g}^\rho(z)|\, dz = a(\mathfrak{g}^\rho, |\mathfrak{g}^\rho|) = a(|\mathfrak{g}^\rho|, \mathfrak{g}^\rho).$$

Hence $a(|\mathfrak{g}^\rho| - \mathfrak{g}^\rho, |\mathfrak{g}^\rho| - \mathfrak{g}^\rho) \leq 0$ and, by the coercivity of a, $\mathfrak{g}^\rho = |\mathfrak{g}^\rho| \geq 0$ a.e. on Ω.

Let $R \geq \delta(y) > 100\rho > 0$. We write again B_R for $B(y,R)$. Let p in the range given by Lemma 4.2, that is $p \in [1, 2n/(n-2)]$ if $n \geq 3$ and $p \in [1,+\infty)$ if $n = 2$. We aim to prove that for all $t > 0$,

$$\frac{m(\{x \in B_R,\, \mathfrak{g}^\rho(x) > t\})}{m(B_R)} \leq C t^{-\frac{p}{2}} R^{\frac{p}{2}(1-d)} \tag{10.21}$$

with a constant C independent of ρ, t and R.

We use (10.16) with the test function

$$\varphi(x) := \left(\frac{2}{t} - \frac{1}{\mathfrak{g}^\rho(x)}\right)^+ = \max\left\{0, \frac{2}{t} - \frac{1}{\mathfrak{g}^\rho(x)}\right\} \tag{10.22}$$

(and $\varphi(x) = 0$ if $\mathfrak{g}^\rho(x) = 0$), which lies in W_0 by Lemma 6.1. So if $\Omega_s := \{x \in \Omega,\ \mathfrak{g}^\rho(x) > s\}$, we have

$$a(\mathfrak{g}^\rho, \varphi) = \int_{\Omega_{t/2}} \frac{\mathcal{A}\nabla \mathfrak{g}^\rho \cdot \nabla \mathfrak{g}^\rho}{(\mathfrak{g}^\rho)^2}\, dm = \fint_{B_r} \varphi \leq \frac{2}{t}. \tag{10.23}$$

Therefore, with the ellipticity condition (8.9),

$$\int_{\Omega_{t/2}} \frac{|\nabla \mathfrak{g}^\rho|^2}{(\mathfrak{g}^\rho)^2}\, dm \leq \frac{C}{t}. \tag{10.24}$$

Pick $y_0 \in \Gamma$ such that $|y - y_0| = \delta(y)$. Set $\widetilde{B}_R$ for $B(y_0, 2R) \supset B_R$. Also define v by $v(x) := (\ln(\mathfrak{g}^\rho(x)) - \ln t + \ln 2)^+$, which lies in W_0 too, thanks to Lemma 6.1. The Sobolev-Poincaré inequality (4.13) implies that

$$\begin{aligned}\left(\int_{\Omega_{t/2}\cap \widetilde{B}_R} |v|^p\, dm\right)^{\frac{1}{p}} &\leq CRm(\widetilde{B}_R)^{\frac{1}{p}-\frac{1}{2}} \left(\int_{\Omega_{t/2}\cap \widetilde{B}_R} |\nabla v|^2\, dm\right)^{\frac{1}{2}} \\ &\leq CRm(\widetilde{B}_R)^{\frac{1}{p}-\frac{1}{2}} t^{-\frac{1}{2}}\end{aligned} \tag{10.25}$$

by (10.24). Since $m(\widetilde{B}_R) \approx R^{d+1}$ thanks to Lemma 2.3, one has

$$\int_{\Omega_{t/2}\cap B_R} \left|\ln\left(\frac{2\mathfrak{g}^\rho}{t}\right)\right|^p dm \leq CR^{p+(d+1)(1-\frac{p}{2})} t^{-\frac{p}{2}}. \tag{10.26}$$

But the latter implies, since $v > \ln 2$ on Ω_t, that

$$(\ln 2)^p m(\Omega_t \cap B_R) \leq CR^{p+(d+1)(1-\frac{p}{2})} t^{-\frac{p}{2}} = Ct^{-\frac{p}{2}} R^{\frac{p}{2}(1-d)+(d+1)}. \tag{10.27}$$

The claim (10.21) follows once we notice that, due to Lemma 2.3, we have $m(B_R) \approx R^{d+1}$.

Now we give a pointwise estimate on $\mathfrak{g}^\rho$ when x is far from y. We claim that

$$\mathfrak{g}^\rho(x) \leq C|x-y|^{1-d} \qquad \text{if } 4|x-y| \geq \delta(y) > 100\rho, \tag{10.28}$$

where again $C > 0$ is independent of ρ. Set $R = 4|x-y| > \delta(y)$. Recall (10.17), i.e., that $\mathfrak{g}^\rho$ lies in W_0 and is a solution in $\Omega \setminus \overline{B_\rho}$. So we can use the Moser estimates to get that

$$\mathfrak{g}^\rho(x) \leq C\frac{1}{m(B(x,R/2))} \int_{B(x,R/2)} \mathfrak{g}^\rho\, dm. \tag{10.29}$$

Indeed, (10.29) is obtained with Lemma 8.7 when $\delta(x) \geq R/30$ (apply Moser inequality in the ball $B(x, R/90)$) and with Lemma 8.13 when $\delta(x) \leq R/30$ (apply Moser inequality in the ball $B(x_0, R/15)$ where x_0 is such that $|x - x_0| = \delta(x)$).

We can use now the fact that $B(x, R/2) \subset B_R$ and [**Duo**, p. 28, Proposition 2.3] to get

$$\mathfrak{g}^\rho(x) \leq C \int_0^{+\infty} \frac{m(\Omega_t \cap B_R)}{m(B_R)} dt \tag{10.30}$$

Take $s > 0$, to be chosen later. By (10.21), applied with any valid $p > 2$ (for instance $p = \frac{2n}{n-1}$),

$$\begin{aligned} \mathfrak{g}^\rho(x) &\leq C\int_0^s \frac{m(\Omega_t \cap B_R)}{m(B_R)} dt + C\int_s^{+\infty} \frac{m(\Omega_t \cap B_R)}{m(B_R)} dt \\ &\leq Cs + CR^{\frac{p}{2}(1-d)} \int_s^{+\infty} t^{-\frac{p}{2}} dt \leq Cs + CR^{\frac{p}{2}(1-d)} s^{1-\frac{p}{2}}. \end{aligned} \tag{10.31}$$

We minimize the right-hand side in s. We find $s \approx R^{1-d}$ and then $\mathfrak{g}^\rho(x) \leq CR^{1-d}$. The claim (10.28) follows.

Let us now prove some pointwise estimates on $\mathfrak{g}^\rho$ when x is close to y. When $n \geq 3$, we want to show that

$$\mathfrak{g}^\rho(x) \leq C\frac{|x-y|^{2-n}}{w(y)} \qquad \text{if } \delta(y) \geq 2|x-y| > 4\rho \text{ and } \delta(y) > 100\rho, \tag{10.32}$$

where $C > 0$ is independent of ρ, x and y. When $n = 2$, we claim that for any $\epsilon > 0$,

$$\mathfrak{g}^\rho(x) \leq C_\epsilon \frac{1}{w(y)} \left(\frac{\delta(y)}{r}\right)^\epsilon \qquad \text{if } \delta(y) \geq 2|x-y| > 4\rho \text{ and } \delta(y) > 100\rho, \tag{10.33}$$

where $C_\epsilon > 0$ is also independent of ρ, x and y. The proof works a little like when x is far from y, but we need to be a bit more careful about the Poincaré-Sobolev inequality that we use. Set again $r = 2|x-y|$. Lemma 8.7 applied to the ball $B(x, r/20)$ yields

$$\mathfrak{g}^\rho(x) \leq \frac{C}{m(B(x,r/2))} \int_{B(x,r/2)} \mathfrak{g}^\rho \, dm \leq \frac{C}{m(B_r)} \int_{B_r} \mathfrak{g}^\rho \, dm \tag{10.34}$$

and then for $s > 0$ and $R > r$ to be chosen soon,

$$\mathfrak{g}^\rho(x) \leq C\int_0^s \frac{m(\Omega_t \cap B_r)}{m(B_r)} dt + C\frac{m(B_R)}{m(B_r)} \int_s^{+\infty} \frac{m(\Omega_t \cap B_R)}{m(B_R)} dt. \tag{10.35}$$

Take $R = \delta(y)$. The doubling property (2.8) allows us to estimate $\frac{m(B_R)}{m(B_r)}$ by $\left(\frac{\delta(y)}{r}\right)^n$. Let p lie in the range given by Lemma 4.2, and apply (10.21) to estimate $m(\Omega_t \cap B_R)$; we get that

$$\frac{m(\Omega_t \cap B_R)}{m(B_R)} \leq Ct^{-p/2} R^{\frac{p}{2}(1-d)} \leq C_p t^{-\frac{p}{2}} \delta(y)^{\frac{p}{2}(1-d)}. \tag{10.36}$$

The bound (10.35) becomes now

$$\mathfrak{g}^\rho(x) \leq Cs + C_p \left(\frac{\delta(y)}{r}\right)^n \delta(y)^{\frac{p}{2}(1-d)} \int_s^{+\infty} t^{-\frac{p}{2}} dt \leq Cs + C_p \delta(y)^{\frac{p}{2}(1-d)+n} r^{-n} s^{1-\frac{p}{2}}. \tag{10.37}$$

We minimize then the right hand side of (10.37) in s. We take $s \approx \delta(y)^{1-d}\left(\frac{\delta(y)}{r}\right)^{\frac{2n}{p}}$ and get that

$$\mathfrak{g}^\rho(x) \leq C_p \delta(y)^{1-d} \left(\frac{\delta(y)}{r}\right)^{\frac{2n}{p}}. \tag{10.38}$$

The assertion (10.32) follows from (10.38) by taking $p = \frac{2n}{n-2}$ (which is possible since $n \geq 3$) and by recalling that $w(y) = \delta(y)^{d+1-n}$. When $n = 2$, we have

$\delta(y)^{1-d} = \delta(y)^{n-d-1} = w(y)^{-1}$ and so (10.33) is obtained from (10.38) by taking $p = \frac{2n}{\epsilon} < +\infty$.

Next we give a bound on the L^q-norm of the gradient of $\mathfrak{g}^\rho$ for some $q > 1$. As before, we want the bound to be independent of ρ so that we can later let our Green function be a weak limit of a subsequence of $\mathfrak{g}^\rho$.

We want to prove first the following Caccioppoli-like inequality: for any $r > 4\rho$,

$$\int_{\Omega \setminus B_r} |\nabla \mathfrak{g}^\rho|^2 \, dm \leq C r^{-2} \int_{B_r \setminus B_{r/2}} (\mathfrak{g}^\rho)^2 dm, \tag{10.39}$$

where $C > 0$ is a constant that depends only upon d, n, C_0 and C_1.

Keep $r > 4\rho$, and let $\alpha \in C^\infty(\mathbb{R}^n)$ be such that $\alpha \equiv 1$ on $\mathbb{R}^n \setminus B_r$, $\alpha \equiv 0$ on $B_{r/2}$ and $|\nabla \alpha| \leq \frac{4}{r}$. By construction, g^ρ lies in W_0, and thus the function $\varphi := \alpha^2 \mathfrak{g}^\rho$ is supported in $\Omega \setminus \overline{B_{r/4}}$ and lies in W_0 thanks to Lemma 5.4. Since we like function with compact support, let us further multiply φ by a smooth, compactly supported function ψ_R such that $\psi_R \equiv 1$ on a large ball B_R. Then $\psi_R \varphi$ is compactly supported in $\Omega \setminus \overline{B_\rho}$, and still lies in W_0 like φ.

Also, (10.17) says that $\mathfrak{g}^\rho$ lies in W_0 and is a solution of $L\mathfrak{g}^\rho = 0$ in $\Omega \setminus \overline{B_\rho} \supset \Omega \setminus \overline{B_{r/4}}$. So we may apply the second item of Lemma 8.3, with $E = \Omega \setminus \overline{B_{r/4}}$, and we get that

$$\int_\Omega \mathcal{A} \nabla \mathfrak{g}^\rho \cdot \nabla(\psi_R \varphi) \, dm = 0, \tag{10.40}$$

but we would prefer to know that

$$\int_\Omega \mathcal{A} \nabla \mathfrak{g}^\rho \cdot \nabla \varphi \, dm = 0. \tag{10.41}$$

Fortunately, we proved in (ii) of Lemma 5.5 that with correctly chosen functions ψ_R, the product $\psi_R \varphi$ tends to φ in W; see (5.32) in particular. Then

$$\begin{aligned} \left| \int_\Omega \mathcal{A} \nabla \mathfrak{g}^\rho \cdot [\nabla \varphi - \nabla(\psi_R \varphi)] \, dm \right| &\leq C \|\nabla \mathfrak{g}^\rho\|_{L^2(dm)} \, \|\nabla \varphi - \nabla(\psi_R \varphi)\|_{L^2(dm)} \\ &\leq C \|\mathfrak{g}^\rho\|_W \, \|\varphi - (\psi_R \varphi)\|_W \end{aligned} \tag{10.42}$$

by the boundedness property (8.8) of $\mathcal{A}$. The right-hand side tends to 0, so (10.41) follows from (10.40). Since $\varphi = \alpha^2 \mathfrak{g}^\rho$, (10.41) yields

$$\int_\Omega \alpha^2 [\mathcal{A} \nabla \mathfrak{g}^\rho \cdot \nabla \mathfrak{g}^\rho] \, dm = -2 \int_\Omega \alpha \mathfrak{g}^\rho [\mathcal{A} \nabla \mathfrak{g}^\rho \cdot \nabla \alpha] \, dm. \tag{10.43}$$

Together with the elliptic and boundedness conditions on $\mathcal{A}$ (see (8.9) and (8.8)) and the Cauchy-Schwarz inequality, (10.43) becomes

$$\begin{aligned} \int_\Omega \alpha^2 |\nabla \mathfrak{g}^\rho|^2 dm &\leq C \int_\Omega \alpha \mathfrak{g}^\rho |\nabla \mathfrak{g}^\rho| |\nabla \alpha| \, dm \\ &\leq C \left(\int_\Omega \alpha^2 |\nabla \mathfrak{g}^\rho|^2 dm \right)^{\frac{1}{2}} \left(\int_\Omega (\mathfrak{g}^\rho)^2 |\nabla \alpha|^2 dm \right)^{\frac{1}{2}}, \end{aligned} \tag{10.44}$$

which can be rewritten

$$\int_\Omega \alpha^2 |\nabla \mathfrak{g}^\rho|^2 dm \leq C \int_\Omega (\mathfrak{g}^\rho)^2 |\nabla \alpha|^2 dm. \tag{10.45}$$

The bound (10.39) is then a straightforward consequence of our choice of α.

Set $\hat{\Omega}_t = \{x \in \Omega, |\nabla \mathfrak{g}^\rho| > t\}$. As before, there will be two different behaviors. We first check that

$$m(\hat{\Omega}_t) \leq C t^{-\frac{d+1}{d}} \qquad \text{when } t \leq \delta(y)^{-d}. \tag{10.46}$$

Let $r \geq \delta(y)$ be given, to be chosen later. The Caccioppoli-like inequality (10.39) and the pointwise bound (10.28) give

$$\int_{\Omega \setminus B_r} |\nabla \mathfrak{g}^\rho|^2 \, dm \leq C r^{-2} \int_{B_r \setminus B_{r/2}} (\mathfrak{g}^\rho)^2 dm \leq C r^{-2d} m(B_r) \leq C r^{1-d} \tag{10.47}$$

by (2.2), and hence

$$m(\hat{\Omega}_t \setminus B_r) \leq C t^{-2} r^{1-d}. \tag{10.48}$$

This yields

$$m(\hat{\Omega}_t) \leq C t^{-2} r^{1-d} + m(B_r) = C t^{-2} r^{1-d} + C r^{1+d} \tag{10.49}$$

because $r \geq \delta(y)$. Take $r = t^{-\frac{1}{d}}$ in (10.49) (and notice that $r \geq \delta(y)$ when $t \leq \delta(y)^{-d}$). The claim (10.46) follows.

We also want a version of (10.46) when t is big. We aim to prove that

$$m(\hat{\Omega}_t) \leq C w(y)^{-\frac{1}{n-1}} t^{-\frac{n}{n-1}} \qquad \text{when } t \geq \delta(y)^{-d} \text{ and } n \geq 3 \tag{10.50}$$

and for any $\eta \in (0,2)$,

$$m(\hat{\Omega}_t) \leq C_\eta w(y)^{-1} \delta(y)^{d\eta} t^{\eta-2} \qquad \text{when } t \geq \delta(y)^{-d} \text{ and } n = 2. \tag{10.51}$$

The proof of (10.50) is similar to (10.46) but has an additional difficulty: we cannot use the Caccioppoli-like argument (10.39) when r is smaller than 4ρ. So we will use another way. By (10.16) for the test function $\phi = \mathfrak{g}^\rho$ and the elliptic condition (8.9),

$$\int_\Omega |\nabla \mathfrak{g}^\rho|^2 dm \leq C \int_\Omega \mathcal{A} \nabla \mathfrak{g}^\rho \cdot \nabla \mathfrak{g}^\rho \, dm = C \fint_{B_\rho} \mathfrak{g}^\rho(z) \, dz \leq \frac{C}{m(B_\rho)} \int_{B_\rho} \mathfrak{g}^\rho \, dm \tag{10.52}$$

by (2.13). Let y_0 be such that $|y - y_0| = \delta(y)$. We use Hölder's inequality, and then the Sobolev-Poincaré inequality (4.13), with p in the range given by Lemma 4.2, to get that

$$\begin{aligned} \int_\Omega |\nabla \mathfrak{g}^\rho|^2 dm &\leq C_p m(B_\rho)^{-1} m(B_\rho)^{1-\frac{1}{p}} \left(\int_{B_\rho} (\mathfrak{g}^\rho)^p \, dm \right)^{\frac{1}{p}} \\ &\leq C_p m(B_\rho)^{-\frac{1}{p}} \left(\int_{B(y_0, 2\delta(y))} (\mathfrak{g}^\rho)^p \, dm \right)^{\frac{1}{p}} \\ &\leq C_p m(B_\rho)^{-\frac{1}{p}} \delta(y) m(B_{3\delta(y)})^{\frac{1}{p}-\frac{1}{2}} \left(\int_\Omega |\nabla \mathfrak{g}^\rho|^2 \, dm \right)^{\frac{1}{2}}, \end{aligned} \tag{10.53}$$

that is,

$$\int_\Omega |\nabla \mathfrak{g}^\rho|^2 dm \leq C_p m(B_\rho)^{-\frac{2}{p}} \delta(y)^2 m(B_{\delta(y)})^{\frac{2}{p}-1}. \tag{10.54}$$

We use the fact that $100\rho < \delta(y)$ and Lemma 2.3 to get that $m(B_\rho) \approx \rho^n w(y) = \rho^n \delta(y)^{d+1-n}$. Besides, notice that $m(B_{3\delta(y)}) \approx \delta(y)^{d+1}$. We end up with

$$(10.55)\quad \int_\Omega |\nabla \mathfrak{g}^\rho|^2 dm \leq C_p \rho^{-\frac{2n}{p}} w(y)^{-\frac{2}{p}} \delta(y)^{2+(d+1)(\frac{2}{p}-1)} = C_p \left(\frac{\delta(y)}{\rho}\right)^{\frac{2n}{p}} \delta(y)^{1-d}$$

once we recall that $w(y) = \delta(y)^{d+1-n}$. Observe that the right-hand side of (10.55) is similar to the one of (10.38). In the same way as below (10.38) we take $p = \frac{2n}{n-2}$ when $n \geq 3$ and $p = \frac{4}{\epsilon}$ when $n = 2$, and obtain that

$$(10.56)\quad \int_\Omega |\nabla \mathfrak{g}^\rho|^2 dm \leq \begin{cases} Cw(y)^{-1}\rho^{2-n} & \text{if } n \geq 3 \\ C_\epsilon w(y)^{-1}\left(\frac{\delta(y)}{\rho}\right)^\epsilon & \text{for any } \epsilon > 0 \text{ if } n = 2. \end{cases}$$

Let $r \leq \delta(y)$, to be chosen soon. Now we show that

$$(10.57)\quad \int_{\Omega \setminus B_r} |\nabla \mathfrak{g}^\rho|^2 dm \leq \begin{cases} Cw(y)^{-1} r^{2-n} & \text{if } n \geq 3 \\ C_\epsilon w(y)^{-1}\left(\frac{\delta(y)}{r}\right)^\epsilon & \text{for any } \epsilon > 0 \text{ if } n = 2. \end{cases}$$

When $r \leq 4\rho$, this is a consequence of (10.56), and when $4\rho < r \leq \delta(y)$, this can be proven as we proved (10.47), by using Caccioppoli-like inequality (10.39) and the pointwise bounds (10.32) or (10.33). That is, we say that

$$(10.58)\quad \begin{aligned} \int_{\Omega\setminus B_r} |\nabla \mathfrak{g}^\rho|^2 dm &\leq Cr^{-2}\int_{B_r\setminus B_{r/2}} (\mathfrak{g}^\rho)^2 dm \\ &\leq r^{-2} m(B_r) \frac{1}{w(y)^2} \begin{cases} Cr^{2(2-n)} & \text{if } n \geq 3 \\ C_\epsilon \left(\frac{\delta(y)}{r}\right)^{2\epsilon} & \text{if } n = 2,\ \epsilon > 0 \end{cases} \end{aligned}$$

and we observe that $m(B_r) \approx w(y) r^n$.

Let $n \geq 3$. We deduce from (10.57) that $m(\hat{\Omega}_t \setminus B_r) \leq Ct^{-2} r^{2-n} w(y)^{-1}$ and then, since $m(B_r) \leq Cr^n w(y)$ and thanks to Lemma 2.3,

$$(10.59)\quad m(\hat{\Omega}_t) \leq Cw(y)^{-1} t^{-2} r^{2-n} + m(B_r) \leq Ct^{-2} w(y)^{-1} r^{2-n} + Cr^n w(y).$$

Choose $r = [tw(y)]^{-\frac{1}{n-1}}$ (which is smaller than $\delta(y)$ if $t \geq \delta(y)^{-d}$) in (10.59). This yields (10.50).

Let $n = 2$ and let $\eta \in (0, 2)$ be given. Set $\epsilon := \frac{2\eta}{2-\eta} > 0$. In this case, (10.57) gives

$$(10.60)\quad m(\hat{\Omega}_t \setminus B_r) \leq Ct^{-2} w(y)^{-1} \left(\frac{\delta(y)}{r}\right)^\epsilon$$

and then since $m(B_r) \leq Cr^2 w(y)$ by Lemma 2.3,

$$(10.61)\quad m(\hat{\Omega}_t) \leq Ct^{-2} w(y)^{-1} \left(\frac{\delta(y)}{r}\right)^\epsilon + Cr^2 w(y).$$

We want to minimize the above quantity in r. We take $r = \delta(y)^{\frac{2(1-d)+\epsilon}{2+\epsilon}} t^{-\frac{2}{2+\epsilon}}$, which is smaller than $\delta(y)$ when $t \geq \delta(y)^{-d}$ and we find that

$$(10.62)\quad m(\hat{\Omega}_t) \leq Ct^{-\frac{4}{2+\epsilon}} \delta(y)^{\frac{2(1-d)+\epsilon(d+1)}{2+\epsilon}} = Ct^{\eta-2}\delta(y)^{1-d+\eta d},$$

with our choice of ϵ. Since $w(y)^{-1} = \delta(y)^{1-d}$ when $n = 2$, the claim (10.51) follows.

We plan to show now that $\nabla \mathfrak{g}^\rho \in L^q(B_R, w)$ for $1 \leq q < n/(n-1)$, and the $L^q(B_R, w)$-norm of $\nabla \mathfrak{g}^\rho$ can be bounded uniformly in ρ. More precisely, we claim that for $R \geq \delta(y)$ and $1 \leq q < n/(n-1)$,

$$\int_{B_R} |\nabla \mathfrak{g}^\rho|^q dm \leq C_q R^{d(1-q)+1}, \tag{10.63}$$

where C_q is independent of ρ and R.

Let $s \in (0, \delta(y)^{-d}]$ be given, to be chosen soon. Then

$$\int_{B_R} |\nabla \mathfrak{g}^\rho|^q dm \leq C \int_0^s t^{q-1} m(B_R) dt + C \int_s^{\delta(y)^{-d}} t^{q-1} m(\hat{\Omega}_t \cap B_R) dt \\ + C \int_{\delta(y)^{-d}}^{+\infty} t^{q-1} m(\hat{\Omega}_t \cap B_R) dt. \tag{10.64}$$

Let us call I_1, I_2 and I_3 the three integrals in the right hand side of (10.64). By Lemma 2.3, $I_1 \leq C s^q m(B_R) \leq C s^q R^{d+1}$. The second integral I_2 is bounded with the help of (10.46), which gives

$$I_2 \leq C \int_s^{\delta(y)^{-d}} t^{q-1-\frac{d+1}{d}} dt \leq C \left(s^{q-\frac{d+1}{d}} - \delta(y)^{d(1-q)+1} \right). \tag{10.65}$$

When $n \geq 3$, the last integral I_3 is bounded with the help of (10.50) and we obtain, when $q < \frac{n}{n-1}$,

$$I_3 \leq C w(y)^{-\frac{1}{n-1}} \int_{\delta(y)^{-d}}^{+\infty} t^{q-1-\frac{n}{n-1}} dt \leq C w(y)^{-\frac{1}{n-1}} \delta(y)^{-qd+\frac{nd}{n-1}} = C\delta(y)^{1+d(1-q)} \tag{10.66}$$

where the last equality is obtained by using the fact that $w(y) = \delta(y)^{d+1-n}$. Note also that the same bound (10.66) can be obtained when $n = 2$ by using (10.51) with $\eta = \frac{2-q}{2}$. The left-hand side of (10.64) can be now bounded for every $n \geq 2$ by

$$\int_{B_R} |\nabla \mathfrak{g}^\rho|^q dm \leq C s^q R^{d+1} + C \left(s^{q-\frac{d+1}{d}} - \delta(y)^{d(1-q)+1} \right) \\ + C\delta(y)^{1+d(1-q)} = C s^q R^{d+1} + C s^{q-\frac{d+1}{d}}, \tag{10.67}$$

where the third term in the middle is dominated by $s^{q-\frac{d+1}{d}}$ because $I_2 \geq 0$. We take $s = R^{-d} \leq \delta(y)^{-d}$ in the right hand side of (10.67) to get the claim (10.63).

As we said, we want to define the Green function as a weak limit of functions $\mathfrak{g}^\rho$, $0 < \rho \leq \delta(y)/100$. We want to prove that for $q \in (1, \frac{n}{n-1})$ and $R > 0$,

$$\|\mathfrak{g}^\rho\|_{W^{1,q}(B_R)} \leq C_{q,R}, \tag{10.68}$$

where $C_{q,R}$ is independent of ρ (but depends, among others things, on y, q and R). First, it is enough to prove the result for $R \geq 2\delta(y)$. Thanks to (10.63), the quantity $\|\nabla \mathfrak{g}^\rho\|_{L^q(B_R, w)}$ is bounded uniformly in $\rho \in (0, \delta(y)/100)$. Due to (2.13), the quantity $\|\nabla \mathfrak{g}^\rho\|_{L^q(B_R)}$ is bounded uniformly in ρ. Now, due to [**Maz**, Corollary 1.1.11], we deduce that $\mathfrak{g}^{\rho_\eta} \in W^{1,q}(B_R)$ and hence with the classical Poincaré inequality on balls that

$$\int_{B_R} \left| \mathfrak{g}^\rho(z) - \fint_{B_R} \mathfrak{g}^\rho(y)\, dy \right|^q dz \leq C_{q,R} \|\nabla \mathfrak{g}^\rho\|^q_{L^q(B_R)} \leq C_{q,R}, \tag{10.69}$$

where $C_{q,R} > 0$ is independent of ρ. Choose $y_0 \in \Gamma$ such that $|y - y_0| = \delta(y_0)$. Note that $B(y_0, \delta(y)/2) \subset B_R$ because $R \geq 2\delta(y)$, so (10.69) implies that

(10.70)
$$\left| \fint_{B(y_0,\delta(y)/2)} \mathfrak{g}^\rho(z) - \fint_{B_R} \mathfrak{g}^\rho(y)\, dy \right|^q dz \leq \int_{B_R} \left| \mathfrak{g}^\rho(z) - \fint_{B_R} \mathfrak{g}^\rho(y)\, dy \right|^q dz \leq C_{q,R}$$

and hence also, by the triangle inequality,

(10.71)
$$\int_{B_R} \left| \mathfrak{g}^\rho(z) - \fint_{B(y_0,\delta(y)/2)} \mathfrak{g}^\rho(y)\, dy \right|^q dz \leq C_{q,R} \int_{B_R} \left| \mathfrak{g}^\rho(z) - \fint_{B_R} \mathfrak{g}^\rho(y)\, dy \right|^q dz.$$

Together with (10.69), we obtain

$$\int_{B_R} |\mathfrak{g}^\rho(z)|^q\, dz \leq C_{q,R} \left(1 + \fint_{B(y_0,\delta(y)/2)} |\mathfrak{g}^\rho(y)|\, dy \right)^q \tag{10.72}$$

and since (10.28) gives that $\fint_{B(y_0,\delta(y)/2)} |\mathfrak{g}^\rho|\, dz \leq C\delta(y)^{1-d}$, the claim (10.68) follows.

Fix $q_0 \in (1, \frac{n}{n-1})$, for instance, take $q_0 = \frac{2n+1}{2n-1}$. Due to (10.68), for all $R > 0$, the functions $(\mathfrak{g}_\rho)_{0<100\rho<\delta(y)}$ are uniformly bounded in $W^{1,q_0}(B_R)$. So a diagonal process allows us to find a sequence $(\rho_\eta)_{\eta \geq 1}$ converging to 0 and a function $\mathfrak{g} \in L^1_{loc}(\mathbb{R}^n)$ such that

$$\mathfrak{g}^{\rho_\eta} \rightharpoonup \mathfrak{g} = g(.,y) \text{ in } W^{1,q_0}(B_R), \text{ for all } R > 0. \tag{10.73}$$

Let $q \in (1, \frac{n}{n-1})$ and $R > 0$. The functions $\mathfrak{g}^{\rho_\eta}$ are uniformly bounded in $W^{1,q}(B_R)$ thanks to (10.68). So we can find a subsequence $\mathfrak{g}^{\rho_{\eta'}}$ of $\mathfrak{g}^{\rho_\eta}$ such that $\mathfrak{g}^{\rho_{\eta'}}$ converges weakly to some function $\mathfrak{g}^{(q,R)} \in W^{1,q}(B_R)$. Yet, by uniqueness of the limit, $\mathfrak{g}$ equals $\mathfrak{g}^{(q,R)}$ almost everywhere in B_R. As a consequence, up to a subsequence (that depends on q and R),

$$\mathfrak{g}^{\rho_\eta} \rightharpoonup \mathfrak{g} = g(.,y) \text{ in } W^{1,q}(B_R). \tag{10.74}$$

The assertion (10.3) follows.

We aim now to prove (10.2), that is

$$(1 - \alpha)\mathfrak{g} \in W_0 \tag{10.75}$$

whenever $\alpha \in C_0^\infty(\mathbb{R}^n)$ satisfies $\alpha \equiv 1$ on B_r for some $r > 0$.

So we choose $\alpha \in C_0^\infty(\mathbb{R}^n)$ and $r > 0$ such that $\alpha \equiv 1$ on B_r. Since α is compactly supported, we can find $R > 0$ such that $\operatorname{supp} \alpha \subset B_R$. For any $\eta \in \mathbb{N}$ such that $4\rho_\eta \leq r$ and $100\rho_\eta < \delta(y)$,

$$\begin{aligned} \|(1-\alpha)\mathfrak{g}^{\rho_\eta}\|_W &\leq \|\mathfrak{g}^{\rho_\eta} \nabla \alpha\|_{L^2(B_R \setminus B_r, w)} + \|(1-\alpha)\nabla g^{\rho_\eta}\|_{L^2(\Omega \setminus B_r, w)} \\ &\leq C_\alpha \sup_{B_R \setminus B_r} g^{\rho_\eta} + C_\alpha \|\nabla g^{\rho_\eta}\|_{L^2(\Omega \setminus B_r, w)}. \end{aligned} \tag{10.76}$$

Thanks to (10.28), (10.32) and (10.33), the term $\sup_{B_R \setminus B_r} g^{\rho_\eta}$ can be bounded by a constant that doesn't depend on η, provided that $\rho_\eta \leq \min(r/4, \delta(y)/100)$. In the same way, (10.57) proves that $\|\nabla g^{\rho_\eta}\|_{L^2(\Omega \setminus B_r, w)}$ can be also bounded by a constant independent of η. As a consequence, for any η satisfying $4\rho_\eta \leq r$,

$$\|(1-\alpha)\mathfrak{g}^{\rho_\eta}\|_W \leq C_\alpha \tag{10.77}$$

where C_α is independent of η. Note also that for η large enough, $(1-\alpha)\mathfrak{g}^{\rho_\eta}$ belongs to W_0 because $\mathfrak{g}^{\rho_\eta} \in W_0$ by construction, and by Lemma 5.4. Therefore, the functions $(1-\alpha)\mathfrak{g}^{\rho_\eta}$, $\eta \in \mathbb{N}$ large, lie in a fixed closed ball of the Hilbert space W_0. So, up to a subsequence, there exists $f_\alpha \in W_0$ such that $(1-\alpha)\mathfrak{g}^{\rho_\eta} \rightharpoonup f_\alpha$ in W_0. By uniqueness of the limit, we have $(1-\alpha)\mathfrak{g} = f_\alpha \in W_0$, that is

$$(1-\alpha)\mathfrak{g}^{\rho_\eta} \rightharpoonup (1-\alpha)\mathfrak{g} \quad \text{in } W_0. \tag{10.78}$$

The claim (10.75) follows.

Observe that (10.75) implies that $\mathfrak{g} \in W_r(\mathbb{R}^n \setminus \{y\})$. Indeed, take $\varphi \in C_0^\infty(\mathbb{R}^n \setminus \{y\})$. We can find $r > 0$ such that $\varphi \equiv 0$ in B_r. Construct now $\alpha \in C_0^\infty(B_r)$ such that $\alpha \equiv 1$ in $B_{r/2}$ and we have

$$\varphi\mathfrak{g} = \varphi[(1-\alpha)\mathfrak{g}] \in W_0 \subset W \tag{10.79}$$

by (10.75) and Lemma 5.4. Hence $\mathfrak{g} \in W_r(\mathbb{R}^n \setminus \{y\})$.

Now we want to prove (10.4). Fix $q \in (1, n/(n-1))$ and a function $\phi \in C_0^\infty(B_{\delta(y)/2})$ such that $\phi \equiv 1$ in $B_{\delta(y)/4}$. Then let φ be any function in $C_0^\infty(\Omega)$. Let us first check that

$$a(\mathfrak{g}, \phi\varphi) := \int_\Omega A\nabla\mathfrak{g} \cdot \nabla[\phi\varphi] dx = \varphi(y) \tag{10.80}$$

and

$$a(\mathfrak{g}, (1-\phi)\varphi) := \int_\Omega A\nabla\mathfrak{g} \cdot \nabla[(1-\phi)\varphi] dx = 0. \tag{10.81}$$

The map $a(., \phi\varphi)$ is a bounded linear functional on $W^{1,q}(B_{\delta(y)/2})$ and thus the weak convergence (in $W^{1,q}(B_R)$) of a subsequence $\mathfrak{g}^{\rho_{\eta'}}$ of $\mathfrak{g}^{\rho_\eta}$ yields

$$a(\mathfrak{g}, \phi\varphi) = \lim_{\eta' \to +\infty} a(\mathfrak{g}^{\rho_{\eta'}}, \phi\varphi) = \lim_{\rho \to 0} \fint_{B(y,\rho)} \phi\varphi \, dx = \varphi(y), \tag{10.82}$$

which is (10.80). Let $\alpha \in C_0^\infty(B_{\delta(y)/4})$ be such that $\alpha \equiv 1$ on $B_{\delta(y)/8}$. The map $a(., (1-\phi)\varphi)$ is bounded on W_0 thus the weak convergence of a subsequence of $(1-\alpha)\mathfrak{g}^{\rho_\eta}$ to $(1-\alpha)\mathfrak{g}$ in W_0 gives

(10.83)
$$\begin{aligned} a(\mathfrak{g}, (1-\phi)\varphi) &= a((1-\alpha)\mathfrak{g}, (1-\phi)\varphi) \\ &= \lim_{\eta' \to +\infty} a((1-\alpha)\mathfrak{g}^{\rho_{\eta'}}, (1-\phi)\varphi) = \lim_{\eta' \to +\infty} a(\mathfrak{g}^{\rho_{\eta'}}, (1-\phi)\varphi) \\ &= \lim_{\rho \to 0} \fint_{B(y,\rho)} (1-\phi)\varphi \, dx = 0. \end{aligned}$$

which is (10.81). The assertion (10.4) now follows from (10.80) and (10.81).

If we use (10.4) for the functions in $C_0^\infty(\Omega \setminus \{y\})$, we immediately obtain that

$$\mathfrak{g} \text{ is a solution of } L\mathfrak{g} = 0 \text{ on } \Omega \setminus \{y\}. \tag{10.84}$$

Assertions (10.5) and (10.7) come from the weak lower semicontinuity of the L^q-norms and the bounds (10.47), (10.57) and (10.63). Notice also that $r^{1-d} \approx \frac{r^{2-n}}{w(y)}$ when r is near $\delta(y)$, so the cut-off between the different cases does not need to be so precise. Let us show (10.6). Let $R > 0$ be a big given number. We have shown that the sequence $\mathfrak{g}^{\rho_\eta}$ is uniformly bounded in $W^{1,q}(B_R)$. Then, by the Rellich-Kondrachov theorem, there exists a subsequence of $\mathfrak{g}^{\rho_\eta}$ that also converges strongly in $L^1(B_R)$ and then another subsequence of $\mathfrak{g}^{\rho_\eta}$ that converges almost everywhere in B_R. The estimates (10.28), (10.32) and (10.33) yield then

$$(10.85)\quad 0 \le \mathfrak{g}(x) \le \begin{cases} C|x-y|^{1-d} & \text{if } 4|x-y| \ge \delta(y) \\ \frac{C|x-y|^{2-n}}{w(y)} & \text{if } 2|x-y| \le \delta(y),\ n \ge 3 \\ \frac{C_\epsilon}{w(y)}\left(\frac{\delta(y)}{|x-y|}\right)^\epsilon & \text{if } 2|x-y| \le \delta(y),\ n = 2, \end{cases} \quad \text{a.e. on } B_R.$$

But by (10.84) $\mathfrak{g}$ is a solution of $L\mathfrak{g} = 0$ on $\Omega\setminus\{y\}$, so it is continuous on $\mathbb{R}^n\setminus\{y\}$ by Lemmas 8.8 and 8.16, and the bounds (10.85) actually hold pointwise in $\Omega\cap B_R\setminus\{y\}$. Since R can be chosen as large as we want, the bounds (10.6) follow.

It remains to check the weak estimates (10.8) and (10.9). Set $q = \frac{2n+1}{2n-1}$, which satisfies $1 < q < \frac{n}{n-1} < \frac{n}{n-2}$. Let $t > 0$ be given ; by the weak lower semicontinuity of the L^q-norm,

$$(10.86)\quad t^q\,\frac{m(\{x \in B_R,\, \mathfrak{g}(x) > t)}{m(B_R)} \le \frac{1}{m(B_R)}\|\mathfrak{g}\|^q_{L^q(B_R,w)} \le \liminf_{\eta\to+\infty} \frac{1}{m(B_R)}\|\mathfrak{g}^{\rho_\eta}\|^q_{L^q(B_R,w)}.$$

Let us use [**Duo**, p. 28, Proposition 2.3]; in the case of (10.8), we could manage otherwise, but we also want to get (10.9) with the same proof. We observe that

$$\begin{aligned} t^q&\frac{m(\{x \in B_R,\, \mathfrak{g}(x) > t\})}{m(B_R)} \\ &\le \liminf_{\eta\to+\infty}\left[\int_0^t s^{q-1}\frac{m(\{x\in B_R, \mathfrak{g}(x) > t,\, \mathfrak{g}^{\rho_\eta} > s)\}}{m(B_R)}ds \right.\\ &\qquad\left. + \int_t^{+\infty} s^{q-1}\frac{m(\{x\in B_R,\, \mathfrak{g}(x) > t,\, \mathfrak{g}^{\rho_\eta} > s\})}{m(B_R)}ds\right] \\ (10.87)\qquad &\le \frac{t^q}{q}\frac{m(\{x\in B_R,\, \mathfrak{g}(x) > t\})}{m(B_R)} \\ &\qquad + \liminf_{\eta\to+\infty}\int_t^{+\infty} s^{q-1}\frac{m(\{x\in B_R, \mathfrak{g}^{\rho_\eta} > s\})}{m(B_R)}ds. \end{aligned}$$

Let p lie in the range given by Lemma 4.2. The bounds (10.21) gives

$$(10.88)\quad \begin{aligned} t^q\frac{m(\{x\in B_R,\, \mathfrak{g}(x) > t\})}{m(B_R)} &\le C\liminf_{\eta\to+\infty}\int_t^{+\infty} s^{q-1}\frac{m(\{x\in B_R, \mathfrak{g}^{\rho_\eta} > s\})}{m(B_R)}ds \\ &\le C_pR^{\frac{p}{2}(1-d)}\int_t^{+\infty} s^{q-1-\frac{p}{2}}ds \le C_pR^{\frac{p}{2}(1-d)}t^{q-\frac{p}{2}}. \end{aligned}$$

The estimates (10.8) follows by dividing both sides of (10.88) by t^q. The same ideas are used to prove (10.9) from (10.46), (10.50) and (10.51). This finally completes the proof of Lemma 10.1. □

LEMMA 10.5. *Any non-negative function $g : \Omega\times\Omega \to \mathbb{R}\cup\{+\infty\}$ that verifies the following conditions:*

(i) *for every $y \in \Omega$ and $\alpha \in C_0^\infty(\mathbb{R}^n)$ such that $\alpha \equiv 1$ in $B(y, r)$ for some $r > 0$, the function $(1-\alpha)g(.,y)$ lies in W_0,*
(ii) *for every $y \in \Omega$, the function $g(.,y)$ lies in $W^{1,1}(B(y,\delta(y)))$,*

(iii) *for* $y \in \Omega$ *and* $\varphi \in C_0^\infty(\Omega)$,

$$\int_\Omega A\nabla_x g(x,y) \cdot \nabla\varphi(x)dx = \varphi(y), \tag{10.89}$$

enjoys the following pointwise lower bound:

(10.90)

$$g(x,y) \geq C^{-1} \frac{|x-y|^2}{m(B(y,|x-y|))} \approx \frac{|x-y|^{2-n}}{w(y)} \quad \textit{for } x,y \in \Omega \textit{ such that } 0 < |x-y| \leq \frac{\delta(y)}{2}.$$

PROOF. Let g satisfy the assumptions of the lemma, fix $y \in \Omega$, write $\mathfrak{g}(x)$ for $g(x,y)$, and use B_r for $B(y,r)$. Thus we want to prove that

$$\mathfrak{g}(x) \geq \frac{|x-y|^2}{Cm(B_{|x-y|})} \qquad \text{whenever } 0 < |x-y| \leq \frac{\delta(y)}{2}. \tag{10.91}$$

With our assumptions, $\mathfrak{g} \in W_r(\mathbb{R}^n \setminus \{y\})$ and it is a solution in $\Omega \setminus \{y\}$ with zero trace; the proof is the same as for (10.79) and (10.84) in Lemma 10.1. Take $x \in \Omega \setminus \{y\}$ such that $|x-y| \leq \frac{\delta(y)}{2}$. Write r for $|x-y|$ and let $\alpha \in C_0^\infty(\Omega \setminus \{y\})$ be such that $\alpha = 1$ on $B_r \setminus B_{r/2}$, $\alpha = 0$ outside of $B_{3r/2} \setminus B_{r/4}$, and $|\nabla\alpha| \leq 8/r$. Using Caccioppoli's inequality (Lemma 8.6) with the cut-off function α, we obtain

(10.92)

$$\begin{aligned}\int_{B_r \setminus B_{r/2}} |\nabla\mathfrak{g}|^2 dm &\leq Cr^{-2} \int_{B_{3r/2}\setminus B_{r/4}} \mathfrak{g}^2 dm \\ &\leq Cr^{-2} m(B_{3r/2}) \sup_{B_{3r/2}\setminus B_{r/4}} \mathfrak{g}^2 \leq Cr^{-2} m(B_r) \sup_{B_{3r/2}\setminus B_{r/4}} \mathfrak{g}^2\end{aligned}$$

by the doubling property (2.8). We can cover $B_{3r/2} \setminus B_{r/4}$ by a finite (independent of y and r) number of balls of radius $r/20$ centered in $B_{3r/2} \setminus B_{r/4}$. Then use the Harnack inequality given by Lemma 8.9 several times, to get that

$$\int_{B_r \setminus B_{r/2}} |\nabla\mathfrak{g}|^2 dm \leq Cr^{-2} m(B_r)\mathfrak{g}(x)^2. \tag{10.93}$$

Define another function $\eta \in C_0^\infty(\Omega)$ which is supported in B_r, equal to 1 on $B_{r/2}$, and such that $|\nabla\eta| \leq \frac{4}{r}$. Use η as a test function in (10.89) to get that

$$1 = \int_{B_r \setminus B_{r/2}} \mathcal{A}\nabla\mathfrak{g} \cdot \nabla\eta \, dm \leq \frac{C}{r} \int_{B_r \setminus B_{r/2}} |\nabla\mathfrak{g}| \, dm, \tag{10.94}$$

where we used (8.8) for the last estimate. Together with the Cauchy-Schwarz inequality and (10.93), this yields

$$1 \leq \frac{C}{r} m(B_r)^{\frac{1}{2}} \Big(\int_{B_r \setminus B_{r/2}} |\nabla\mathfrak{g}|^2 \, dm \Big)^{\frac{1}{2}} \leq Cr^{-2} m(B_r)\mathfrak{g}(x). \tag{10.95}$$

The lower bound (10.91) follows. □

In the sequel, A^T denotes the transpose matrix of A, defined by $A^T_{ij}(x) = A_{ji}(x)$ for $x \in \Omega$ and $1 \leq i,j \leq n$. Thus A^T satisfies the same boundedness and elliptic conditions as A. That is, it satisfies (8.6) and (8.7) with the same constant C_1. We can thus define solutions to $L_T u := -\operatorname{div} A^T \nabla u = 0$ for which the results given in Section 8 hold.

Denote by $g : \Omega \times \Omega \to \mathbb{R} \cup \{+\infty\}$ the Green function defined in Lemma 10.1, and by $g^T : \Omega \times \Omega \to \mathbb{R} \cup \{+\infty\}$ the Green function defined in Lemma 10.1, but with A is replaced by A^T.

LEMMA 10.6. *With the notation above,*

$$g(x,y) = g_T(y,x) \quad \text{for } x,y \in \Omega,\ x \neq y. \tag{10.96}$$

In particular, the functions $y \to g(x,y)$ satisfy the estimates given in Lemma 10.1 and Lemma 10.5.

PROOF. The proof is the same as for [**GW**, Theorem 1.3]. Let us review it for completeness. Let $x, y \in \Omega$ be such that $x \neq y$. Set $B = B(\frac{x+y}{2}, |x-y|)$ and let $q \in (1, \frac{n}{n-1})$.

From the construction given in Lemma 10.1 (see (10.74) in particular), there exists two sequences $(\rho_\nu)_\nu$ and $(\sigma_\mu)_\mu$ converging to 0 such that $g^{\rho_\nu}(.,y)$ and $g_T^{\sigma_\mu}(.,x)$ converge weakly in $W^{1,q}(B)$ to $g(.,y)$ and $g_T(.,x)$ respectively. So, up to additional subsequence extractions, $g^{\rho_\nu}(.,y)$ and $g_T^{\sigma_\mu}(.,x)$ converge to $g(.,y)$ and $g_T(.,x)$, strongly in $L^1(B)$, and then pointwise a.e. in B.

Inserting them as test functions in (10.16), we obtain

$$\int_\Omega A\nabla g^{\rho_\nu}(z,y)\cdot\nabla g_T^{\sigma_\mu}(z,x)dz = \fint_{B(y,\rho_\nu)} g_T^{\sigma_\mu}(z,x)dz = \fint_{B(x,\sigma_\mu)} g^{\rho_\nu}(z,y)dz. \tag{10.97}$$

We want to let σ_μ tend to 0. The term $\fint_{B(y,\rho_\nu)} g_T^{\sigma_\mu}(z,x)dz$ tends to $\fint_{B(y,\rho_\nu)} g_T(z,x)dz$ because $g_T(.,x)^{\sigma_\mu}$ tends to $g_T(.,x)$ in $L^1(B)$. When ρ_ν is small enough, the function $g^{\rho_\nu}(.,y)$ is a solution of $L\mathfrak{g}^{\rho_\nu} = 0$ in $\Omega \setminus \overline{B(y,\rho_\nu)} \ni x$, so it is continuous at x thanks to Lemma 8.8. Therefore, the term $\fint_{B(x,\sigma_\mu)} g^{\rho_\nu}(z,y)dz$ tends to $g^{\rho_\nu}(x,y)$. We deduce, when ν is big enough so that $\rho_\nu < |x-y|$,

$$\fint_{B(y,\rho_\nu)} g_T(z,x)dz = g^{\rho_\nu}(x,y). \tag{10.98}$$

Now let ρ_ν tend to 0 in (10.98). The function $g_T(.,x)$ is a solution for L_T in $\Omega\setminus\{x\}$, so it is continuous on $B(y,\rho_\nu)$ for ν large. Hence the left-hand side of (10.98) converges to $g_T(y,x)$. Thanks to Lemma 8.8, the functions $g^{\rho_\nu}(.,y)$ are uniformly Hölder continuous, so the a.e. pointwise convergence of $g^{\rho_\nu}(.,y)$ to $g(.,y)$ can be improved into a uniform convergence on $B(x, \frac{1}{3}|x-y|)$. In particular $g^{\rho_\nu}(x,y)$ tends to $g(x,y)$ when ρ_ν goes to 0. We get that $g_T(y,x) = g(x,y)$, which is the desired conclusion. □

LEMMA 10.7. *Let $g : \Omega \times \Omega \to \mathbb{R} \cup \{+\infty\}$ be the non-negative function constructed in Lemma 10.1. Then for any $f \in C_0^\infty(\Omega)$, the function u defined by*

$$u(x) = \int g(x,y)f(y)dy \tag{10.99}$$

belongs to W_0 and is a solution of $Lu = f$ in the sense that

$$\int_\Omega A\nabla u\cdot\nabla\varphi\, dx = \int_\Omega \mathcal{A}\nabla u\cdot\nabla\varphi\, dm = \int_\Omega f\varphi \quad \text{for every } \varphi \in W_0. \tag{10.100}$$

PROOF. First, let us check that (10.99) make sense. Since $f \in C_0^\infty(\Omega)$, there exists a big ball B with center y and radius $R > \delta(y)$ such that $\operatorname{supp} f \subset B$. By (10.3) and (10.96), $g(x,.)$ lies in $L^1(B)$. Hence the integral in (10.99) is well defined.

Let $f \in C_0^\infty(\Omega)$. Choose a big ball B_f centered on Γ such that $\text{supp} f \subset B_f$. For any $\varphi \in W_0$,

$$\int_\Omega f\varphi\, dz \leq \|f\|_\infty \int_{B_f} |\varphi|\, dz \leq C_f \|\varphi\|_W \tag{10.101}$$

by Lemma 4.1. So the map $\varphi \in W_0 \to \int f\varphi$ is a bounded linear functional on W_0. Since the map $a(u,v) = \int_\Omega \mathcal{A}\nabla u \cdot \nabla v\, dm$ is bounded and coercive on W_0, the Lax-Milgram theorem yields the existence of $u \in W_0$ such that for any $\varphi \in W_0$,

$$\int_\Omega A\nabla u \cdot \nabla\varphi\, dz = \int_\Omega f\varphi\, dz. \tag{10.102}$$

We want now to show that $u(x) = \int_\Omega g(x,y) f(y) dy$. A key point of the proof uses the continuity of u, a property that we assume for the moment and will prove later on. For every $\rho > 0$, let $g_T^\rho(.,x) \in W_0$ be the function satisfying

$$\int_\Omega A^T \nabla_y g_T^\rho(y,x) \cdot \nabla\varphi(y) dy = \fint_{B(x,\rho)} \varphi(y)\, dy \qquad \text{for every } \varphi \in W_0. \tag{10.103}$$

We use $g_T^\rho(.,x)$ as a test function in (10.102) and get that

(10.104)
$$\begin{aligned}\int_\Omega f(y) g_T^\rho(y,x) dy &= \int_\Omega A\nabla u(y) \cdot \nabla_y g_T^\rho(y,x) dy = \int_\Omega A^T \nabla_y g_T^\rho(y,x) \cdot \nabla u(y) dy \\ &= \fint_{B(x,\rho)} u(y)\, dy,\end{aligned}$$

by (10.103). We take a limit as ρ goes to 0. The right-hand side converges to $u(x)$ because, as we assumed, u is continuous. Choose $R \geq \delta(x)$ so big that $\text{supp} f \subset B(x,R)$, and choose also $q \in (1, \frac{n}{n-1})$. According to (10.74), there exists a sequence ρ_ν converging to 0 such that $g_T^{\rho_\nu}(.,x)$ converges weakly in $W^{1,q}(B(x,R)) \subset L^1(B(x,R))$ to the function $g_T(.,x)$, the latter being equal to $g(x,.)$ by Lemma 10.6. Hence

$$\lim_{\nu\to+\infty} \int_\Omega f(y) g_T^{\rho_\nu}(y,x) dy = \int_\Omega f(y) g(x,y) dy \tag{10.105}$$

and then (10.99) holds.

It remains to check what we assumed, that is the continuity of u on Ω. The quickest way to show it is to prove a version of the Hölder continuity (Lemma 8.8) when u is a solution of $Lu = f$. As for the proof of Lemma 8.8, since we are only interested in the continuity inside the domain, we can use the standard elliptic theory, where the result is well known (see for instance [**GT**, Theorem 8.22]). □

The following Lemma states the uniqueness of the Green function.

LEMMA 10.8. *There exists a unique function $g : \Omega \times \Omega \mapsto \mathbb{R} \cup \{+\infty\}$ such that $g(x,.)$ is continuous on $\Omega \setminus \{x\}$ and locally integrable in Ω for every $x \in \Omega$, and such that for every $f \in C_0^\infty(\Omega)$ the function u given by*

$$u(x) := \int_\Omega g(x,y) f(y) dy \tag{10.106}$$

belongs to W_0 and is a solution of $Lu = f$ in the sense that

$$\int_\Omega A\nabla u \cdot \nabla\varphi\, dx = \int_\Omega \mathcal{A}\nabla u \cdot \nabla\varphi\, dm = \int_\Omega f\varphi\, dx \qquad \textit{for every } \varphi \in W_0. \tag{10.107}$$

PROOF. The existence of the Green function is given by Lemma 10.1, Lemma 10.6 and Lemma 10.7. Indeed, if g is the function built in Lemma 10.1, the property (10.3) (together with Lemma 10.6) states that $g(x,.)$ is locally integrable in Ω. The property (10.4) (and Lemma 10.6 again) gives that $g(x,.)$ is a solution in $\Omega \setminus \{x\}$, and thus, by Lemma 8.8, that $g(x,.)$ is continuous in $\Omega \setminus \{x\}$. The last property, i.e. that fact that u given by (10.106) is in W_0 and satisfies (10.107), is exactly Lemma 10.7.

So it remains to prove the uniqueness. Assume that $\tilde{g}$ is another function satisfying the given properties. Thus for $f \in C_0^\infty(\Omega)$, the function $\tilde{u}$ given by

$$\tilde{u}(x) := \int_\Omega \tilde{g}(x,y) f(y) dy \tag{10.108}$$

belongs to W_0 and satisfies $L\tilde{u} = f$. By the uniqueness of the solution of the Dirichlet problem (9.3) (see Lemma 9.1), we must have $\tilde{u} = u$. Therefore, for all $x \in \Omega$ and all $f \in C_0^\infty(\Omega)$,

$$\int_\Omega [\tilde{g}(x,y) - g(x,y)] f(y) dy = 0. \tag{10.109}$$

From the continuity of $g(x,.)$ and $\tilde{g}(x,.)$ in $\Omega \setminus \{x\}$, we deduce that $g(x,y) = \tilde{g}(x,y)$ for any $x, y \in \Omega$, $x \neq y$. $\square$

We end this section with an additional property of the Green function, its decay near the boundary. This property is proven in [**GW**] under the assumption that Ω is of 'S class', which means that we can find an exterior cone at any point of the boundary. We still can prove it in our context because the property relies on the Hölder continuity of solutions at the boundary, that holds in our context because we have (Harnack tubes and) Lemma 8.16.

LEMMA 10.9. *The Green function satisfies*

$$g(x,y) \leq C\delta(x)^\alpha |x-y|^{1-d-\alpha} \quad \textit{for } x, y \in \Omega \textit{ such that } |x-y| \geq 4\delta(x), \tag{10.110}$$

where $C > 0$ and $\alpha > 0$ depend only on n, d, C_0 and C_1.

PROOF. Let $y \in \Omega$ be given. For any $x \in \Omega$, we write $\mathfrak{g}(x)$ for $g(x,y)$. We want to prove that

$$\mathfrak{g}(x) \leq C\delta(x)^\alpha |x-y|^{1-d-\alpha} \quad \text{for } x \in \Omega \text{ such that } |x-y| \geq 4\delta(x), \tag{10.111}$$

with constants $C > 0$ and $\alpha > 0$ that depend only on n, d, C_0 and C_1. By Lemma 10.1-(v),

$$\mathfrak{g}(z) \leq C|z-y|^{1-d} \quad \text{for } z \in \Omega \setminus B(y, \delta(y)/4). \tag{10.112}$$

Let x be such that $|x-y| \geq 4\delta(x)$, choose $x_0 \in \Gamma$ such that $|x - x_0| = \delta(x)$, and set $r = |x-y|$ and $B = B(x_0, |x-y|/3)$; thus $x \in B$. We shall need to know that

$$\delta(y) \leq \delta(x) + |x-y| \leq \frac{r}{4} + r = \frac{5r}{4}. \tag{10.113}$$

Then let z be any point of $\Omega \cap B$. Obviously $|z-x| \leq |z-x_0| + |x_0 - x| \leq \frac{r}{3} + \delta(x) \leq \frac{r}{3} + \frac{r}{4} = \frac{7r}{12}$, which implies that

$$|y-z| \geq |y-x| - |z-x| \geq r - \frac{7r}{12} = \frac{5r}{12} \geq \frac{\delta(y)}{3}. \tag{10.114}$$

Hence by (10.112), $\mathfrak{g}(z) \leq C|z-y|^{1-d}$. Notice also that $|y-z| \leq |y-x|+|z-x| \leq r + \frac{7r}{12} = \frac{19r}{12}$, so, with (10.114), $\frac{5r}{12} \leq |y-z| \leq \frac{19r}{12}$ and

$$\mathfrak{g}(z) \leq Cr^{1-d} = C|x-y|^{1-d} \text{ for } z \in \Omega \cap B, \tag{10.115}$$

even if $d < 1$, and with a constant $C > 0$ that does not dependent on x, y, or x_0.

We now use the fact that $\mathfrak{g}$ is a solution of $L\mathfrak{g} = 0$ on $\Omega \cap B$. Notice that its oscillation on B is the same as its supremum, because it is nonnegative and, by (i) of Lemma 10.5, its trace on $\Gamma \cap B$ vanishes. Lemma 8.16 (the Hölder continuity of solutions at the boundary) says that for some $\alpha > 0$, that depends only on n, d, C_0 and C_1,

(10.116)

$$\begin{aligned} g(x,y) = \mathfrak{g}(x) &\leq \sup_{\overline{B}(x_0,\delta(x))} \mathfrak{g} = \operatorname*{osc}_{\overline{B}(x_0,\delta(x))} \mathfrak{g} \leq C\left(\frac{3\delta(x)}{|x-y|}\right)^{\alpha} \operatorname*{osc}_{B(x_0,|x-y|/3)} \mathfrak{g} \\ &= C\left(\frac{3\delta(x)}{|x-y|}\right)^{\alpha} \sup_{B(x_0,|x-y|/3)} \mathfrak{g} \leq C\left(\frac{\delta(x)}{|x-y|}\right)^{\alpha} |x-y|^{1-d}. \end{aligned}$$

because $B = B(x_0, |x-y|/3)$. The lemma follows. □

CHAPTER 11

The Comparison Principle

In this chapter, we prove two versions of the comparison principle: one for the harmonic measure (Lemma 11.16) and one for locally defined solutions (Lemma 11.17). A big technical difference is that the former is a globally defined solution, while the latter is local.

At the moment we write this manuscript, the proofs of the comparison principle in codimension 1 that we are aware of cannot be straightforwardly adapted to the case of higher codimension. To be more precise, we can indeed prove the comparison principle (in higher codimension) for harmonic measures on Γ by only slightly modifying the arguments of [**CFMS, Ken**]. However, the proof of the comparison principle for solutions (of $Lu = 0$) defined on a *subset* D of Ω in the case of codimension 1 relies on the use of the harmonic measure on the boundary ∂D (see for instance [**CFMS, Ken**]). In our setting, in the case where the considered functions are non-negative and solutions to $Lu = 0$ only on a subset $D \subsetneq \Omega$, we are lacking a definition for harmonic measures with mixed boundaries (some parts in codimension 1 and some parts in higher codimension). The reader can imagine a ball B centered at a point of $\partial\Omega = \Gamma$. The boundary of the $B \cap \Omega$ consists of $\Gamma \cap B$ and ∂B, the sets of different co-dimension. For those reasons, our proof of the comparison principle (in higher codimension) for locally defined functions nontrivially differs from the one in [**CFMS, Ken**]. Therefore, in a first subsection, we illustrate our arguments in the case of codimension 1 to build reader's intuition.

11.1. Discussion of the comparison theorem in codimension 1

We present here two proofs of the comparison principle in the codimension 1 case. The first proof of the one we can find in [**CFMS, Ken**] and the second one is our alternative proof. We consider in this subsection that the reader knows or is able to see the results in the three first sections of [**Ken**], that contain the analogue in codimension 1 of the results proved in the previous sections.

For simplicity, the domain $\Omega \subset \mathbb{R}^n$ that we study is a special Lipschitz domain, that is

$$\Omega = \{(y,t) \in \mathbb{R}^{n-1} \times \mathbb{R},\, \varphi(y) < t\}$$

where $\varphi : \mathbb{R}^{n-1} \to \mathbb{R}$ is a Lipschitz function. The elliptic operator that we consider is $L = -\operatorname{div} A\nabla$, where A is a matrix with bounded measurable coefficients satisfying the classical elliptic condition (see for instance (1.1.1) in [**Ken**]). Yet, the change of variable $\rho : \mathbb{R}^{n-1} \times \mathbb{R} \to \mathbb{R}^{n-1} \times \mathbb{R}$ defined by

$$\rho(y,t) = (y, t - \varphi(y))$$

maps Ω into $\widetilde{\Omega} = \mathbb{R}^n_+ := \{(y,t) \in \mathbb{R}^{n-1} \times \mathbb{R},\, t > 0\}$ and changes the elliptic operator L into $\widetilde{L} = -\operatorname{div} \widetilde{A}\nabla$, where $\widetilde{A}$ is also a matrix with bounded measurable coefficients satisfying the elliptic condition (1.1.1) in [**Ken**]. Therefore, in the sequel, we reduce our choices of Ω and $\Gamma = \partial\Omega$ to $\mathbb{R}^n_+$ and $\mathbb{R}^{n-1} = \{(y,0) \in \mathbb{R}^n,\, y \in \mathbb{R}^{n-1}\}$ respectively.

Let us recall some facts that also hold in the present context. If $u \in W^{1,2}(D)$ and D is a Lipschitz set, then u has a trace on the boundary of D and hence we can give a sense to the expression $u = h$ on ∂D. If in addition, the function u is a solution of $Lu = 0$ in D (the notion of solution is taken in the weak sense, see for instance [**Ken**, Definition 1.1.4]) and h is continuous on ∂D, then u is continuous on $\overline{D}$.

The Green function (associated to the domain $\Omega = \mathbb{R}^n_+$ and the elliptic operator L) is denoted by $g(X,Y)$ - with $X, Y \in \Omega$ - and the harmonic measure (associated to Ω and L) is written $\omega^X(E)$ - with $X \in \Omega$ and $E \subset \Gamma$. The notation $\omega^X_D(E)$ - where $X \in D$ and $E \subset \partial D$ - denotes the harmonic measure associated to the domain D (and the operator L).

When $x_0 = (y_0, 0) \in \Gamma$ and $r > 0$, we use the notation $A_r(x_0)$ for (y_0, r).

In this context, the comparison principle given in [**Ken**, Lemma 1.3.7] is

LEMMA 11.1 (Comparison principle, codimension 1). *Let $x_0 \in \Gamma$ and $r > 0$. Let $u, v \in W^{1,2}(\Omega \cap B(x_0, 2r))$ be two non-negative solutions of $Lu = Lv = 0$ in $\Omega \cap B(x_0, 2r)$ satisfying $u = v = 0$ on $\Gamma \cap B(x_0, 2r)$. Then for any $X \in \Omega \cap B(x_0, r)$, we have*

$$C^{-1}\frac{u(A_r(x_0))}{v(A_r(x_0))} \leq \frac{u(X)}{v(X)} \leq C\frac{u(A_r(x_0))}{v(A_r(x_0))}, \tag{11.1}$$

where $C > 0$ depends only on the dimension n and the ellipticity constants of the matrix A.

PROOF. We recall quickly the ideas of the proof of the comparison principle found in [**Ken**].

Let $x_0 \in \Gamma$ and $r > 0$ be given. We denote $A_r(x_0)$ by X_0 and, for $\alpha > 0$, $B(x_0, \alpha r)$ by B_α.

The proof of (11.1) is reduced to the proof of the upper bound

$$\frac{u(X)}{v(X)} \leq C\frac{u(X_0)}{v(X_0)} \quad \text{for } X \in B_1 \cap \Omega \tag{11.2}$$

because of the symmetry of the role of u and v.

Step 1: Upper bound on u.

By definition of the harmonic measure,

$$u(X) = \int_{\partial(\Omega \cap B_{3/2})} u(y)\, d\omega^X_{\Omega \cap B_{3/2}}(y) \quad \text{for } X \in B_{3/2} \cap \Omega. \tag{11.3}$$

Note that $\partial(\Omega \cap B_{3/2}) = (\partial B_{3/2} \cap \Omega) \cup (\Gamma \cap B_{3/2})$. Hence, for any $X \in B_{3/2} \cap \Omega$,

$$\begin{aligned} u(X) &= \int_{\partial B_{3/2} \cap \Omega} u(y)\, d\omega^X_{\Omega \cap B_{3/2}}(y) + \int_{\Gamma \cap B_{3/2}} u(y)\, d\omega^X_{\Omega \cap B_{3/2}}(y) \\ &= \int_{\partial B_{3/2} \cap \Omega} u(y)\, d\omega^X_{\Omega \cap B_{3/2}}(y) \end{aligned} \tag{11.4}$$

because, by assumption, $u = 0$ on $\Gamma \cap B_2$. Lemma 1.3.5 in [**Ken**] gives now, for any $Y \in B_{7/4}$, the bound $u(Y) \leq Cu(X_0)$ with a constant $C > 0$ which is independent of Y. So by the positivity of the harmonic measure, we have for any $X \in B_{3/2} \cap \Omega$

$$u(X) \leq Cu(X_0) \int_{\partial B_{3/2} \cap \Omega} d\omega^X_{\Omega \cap B_{3/2}}(y) \leq Cu(X_0) \omega^X_{\Omega \cap B_{3/2}}(\partial B_{3/2} \cap \Omega). \tag{11.5}$$

Step 2: Lower bound on v.

First, again by definition of the harmonic measure, we have that for $X \in B_{3/2} \cap \Omega$,

$$v(X) = \int_{\partial(\Omega \cap B_{3/2})} v(y) d\omega^X_{\Omega \cap B_{3/2}}(y). \tag{11.6}$$

Set $E = \{y \in \partial B_{3/2} \cap \Omega;\ \operatorname{dist}(y, \Gamma) \geq \frac{1}{2} r\}$. By assumption, $v \geq 0$ on $\partial(\Omega \cap B_{3/2})$. In addition, thanks to the Harnack inequality, $v(y) \geq C^{-1} v(X_0)$ for every $y \in E$, with a constant $C > 0$ that is independent of y. So the positivity of the harmonic measure yields, for $X \in B_{3/2} \cap \Omega$,

$$v(X) \geq C^{-1} v(X_0) \int_E d\omega^X_{\Omega \cap B_{3/2}}(y) \geq C^{-1} v(X_0) \omega^X_{\Omega \cap B_{3/2}}(E). \tag{11.7}$$

Step 3: Conclusion.

From steps 1 and 2, we deduce that

$$\frac{u(X)}{v(X)} \leq C \frac{u(X_0)}{v(X_0)} \frac{\omega^X_{\Omega \cap B_{3/2}}(\partial B_{3/2} \cap \Omega)}{\omega^X_{\Omega \cap B_{3/2}}(E)} \quad \text{for } X \in \Omega \cap B_{3/2}. \tag{11.8}$$

The inequality (11.2) is now a consequence of the doubling property of the harmonic measure (see for instance (1.3.7) in [**Ken**]), that gives

$$\omega^X_{\Omega \cap B_{3/2}}(\partial B_{3/2} \cap \Omega) \leq C \omega^X_{\Omega \cap B_{3/2}}(E) \quad \text{for } X \in \Omega \cap B_1. \tag{11.9}$$

The lemma follows. □

The proof above relies on the use of the harmonic measure for the domain $\Omega \cap B_{3/2}$. We want to avoid this, and use only the Green functions and harmonic measures related to the domain Ω itself.

First, we need a way to compare two functions in a domain, that is a suitable maximum principle. In the previous proof of Lemma 11.1, the maximum principle was replaced/hidden by the positivity of the harmonic measure, whose proof makes a crucial use of the maximum principle for solutions. See [**Ken**, Definition 1.2.6] for the construction of the harmonic measure, and [**Ken**, Corollary 1.1.18] for the maximum principle. The maximum principle that we will use is the following.

LEMMA 11.2. *Let $F \subset E \subset \mathbb{R}^n$ be two sets such that F is closed, E is open, and $\operatorname{dist}(F, \mathbb{R}^n \setminus E) > 0$. Let u be a solution in $E \cap \Omega$ such that*

(i) $\int_E |\nabla u|^2 \, dx < +\infty$,

(ii) $u \geq 0$ *on* $\Gamma \cap E$,

(iii) $u \geq 0$ *in* $(E \setminus F) \cap \Omega$.

Then $u \geq 0$ in $E \cap \Omega$.

In a more 'classical' maximum principle, assumption (iii) would be replaced by

(iii') $u \geq 0$ in $\partial E \cap \Omega$.

Since this subsection aims to illustrate what we will do in the next subsection, we state here a maximum principle which is as close as possible to the one we will actually prove in higher codimension. Let us mention that using (iii) instead of (iii') will not make computations harder or easier. However, (iii) is much easier to define and use in the higher codimension case (to the point that we did not even try to give a precise meaning to (iii')).

We do not prove Lemma 11.2 here, because the proof is the same as for Lemma 11.3 below, which is its higher codimension version.

Notice that Lemma 11.2 is really a maximum principle where we use the values of u on a boundary $(\Gamma \cap E) \cup (\Omega \cap F \setminus E)$ that surrounds E to control the values of u in $\Omega \cap E$, but here the boundary also has a thick part, $\Omega \cap F \setminus E$. This makes it easier to define Dirichlet conditions on that thick set, which is the main point of (iii).

The first assumption (i) is a technical hypothesis, it can be seen as a way to control u at infinity, which is needed because we actually do not require E or even F to be bounded.

Lemma 11.2 will be used in different situations. For instance, we will use it when $E = 2B$ and $F = B$, where B is a ball centered on Γ.

Step 1 (modified): We want to find an upper estimate for u that avoids using the measure $\omega^X_{\Omega \cap B_{3/2}}$. Lemma 1.3.5 in [**Ken**] gives, as before, that $u(X) \leq Cu(X_0)$ for any $X \in B_{7/4} \cap \Omega$. The following result states the non-degeneracy of the harmonic measure.

$$\omega^X(\Gamma \setminus B_{5/4}) \geq C^{-1} \quad \text{for } X \in \Omega \setminus B_{3/2}, \tag{11.10}$$

where $C > 0$ is independent of x_0, r or X. Indeed, when $X \in \Omega \setminus B_{3/2}$ is close to the boundary, the lower bound (11.10) can be seen as a consequence of the Hölder continuity of solutions. The proof for all $X \in \Omega \setminus B_{3/2}$ is then obtained with the Harnack inequality. See [**Ken**, Lemma 1.3.2] or Lemma 11.10 below for the proof.

From there, we deduce that

$$u(X) \leq Cu(X_0)\omega^X(\Gamma \setminus B_{5/4}) \quad \text{for } X \in \Omega \cap [B_{7/4} \setminus B_{3/2}]. \tag{11.11}$$

We want to use the maximum principle given above (Lemma 11.2), with $E = B_{7/4}$ and $F = B_{3/2}$. However, the function $X \to \omega^X(\Gamma \setminus B_{5/4})$ doesn't satisfy the assumption (i) of Lemma 11.2. So we take $h \in C^\infty(\mathbb{R}^n)$ such that $0 \leq h \leq 1$, $h \equiv 1$ on $\mathbb{R}^n \setminus B_{5/4}$, and $h \equiv 0$ on $\mathbb{R}^n \setminus B_{9/8}$. Define u_h as the only solution of $Lu_h = 0$ in Ω with the Dirichlet condition $u_h = h$ on Γ. We have $u_h(X) \geq \omega^X(\Gamma \setminus B_{5/4})$ by the positivity of the harmonic measure, and thus the bound (11.11) yields the existence of $K_0 > 0$ (independent of x_0, r, X) such that

$$u_1(X) := K_0 u(X_0) u_h(X) - u(X) \geq 0 \quad \text{for } X \in \Omega \cap [B_{7/4} \setminus B_{3/2}]. \tag{11.12}$$

It would be easy to check that $u_1 \geq 0$ on $\Gamma \cap B_{7/4}$ and $\int_E |\nabla u_1| < +\infty$, but we leave the details because they will be done in the larger codimension case. So Lemma 11.2 gives that $u_1 \geq 0$ in $\Omega \cap B_{7/4}$, that is

$$u(X) \leq K_0 u(X_0) u_h(X) \leq K_0 u(X_0) \omega^X(\Gamma \setminus B_{9/8}) \quad \text{for } X \in \Omega \cap B_{7/4}, \tag{11.13}$$

by definition of h and positivity of the harmonic measure.

Step 2 (modified): In the same way, we want to adapt Step 2 of the proof of Lemma 11.1. If we want to proceed as in Step 1, we would like to find and use a function f that keeps the main properties of the object $\omega^X_{\Omega\cap B_{7/4}}(E)$, where $E = \{y \in \partial B_{7/4},\ ;\ \mathrm{dist}(y,\Gamma) \geq r/2\}$. For instance, f such that

(a) f is a solution of $Lf = 0$ in $\Omega \cap B_{7/4}$,
(b) $f \leq 0$ in $\Gamma \cap B_{7/4}$,
(c) $f \leq 0$ in $\{X \in \Omega,\ \mathrm{dist}(X,\Gamma) < r/2\} \cap [B_{7/4} \setminus B_{3/2}]$,
(d) $f(X) \approx \omega^X(\Gamma \setminus B_{9/8})$ in $\Omega \cap B_1$, in particular $f > 0$ in $\Omega \cap B_1$.

The last point is important to be able to conclude (in Step 3). It is given by the doubling property of the harmonic measure (11.9) in the previous proof of Lemma 11.1.

We were not able to find such a function f. However, we can construct an f that satisfies some conditions close to (a), (b), (c) and (d) above. Since f fails to verify exactly (a), (b), (c) and (d), extra computations are needed.

First, note that it is enough to prove that there exists $M > 0$ depending only on n and the ellipticity constants of A, such that for $y_0 \in \Gamma$, $s > 0$, and any non-negative solution v to $Lv = 0$ in $B(y_0, Ms)$

$$v(X) \geq C^{-1} v(A_s(y))\omega^X(\Gamma \setminus B(y_0, 2s)) \quad \text{for } X \in \Omega \cap B(y_0, s), \tag{11.14}$$

where here the the corkscrew point $A_s(y)$ is just $A_s(y) = (y, s)$. Indeed, if we have (11.14), then we can prove that, in the situation of Step 2,

$$v(X) \geq C^{-1} v(X_0)\omega^X(\Gamma \setminus B_2) \quad \text{for } X \in \Omega \cap B_1 \tag{11.15}$$

by using a proper covering of the domain $\Omega \cap B_1$ (if $X \in \Omega \cap B_1$ lies within $\frac{1}{4M}$ of Γ, say, we use (11.14) with $y_0 \in \Gamma$ close to X and $s = \frac{1}{2M}$, and then the Harnack inequality; if instead $X \in \Omega \cap B_1$ is far from the boundary Γ, (11.15) is only a consequence of the Harnack inequality).

The conclusion (11.2) comes then from (11.13), (11.15) and the doubling property of the harmonic measure (see for instance (1.3.7) in [**Ken**]).

It remains to prove the claim (11.14). Let $y_0 \in \Gamma$, $s > 0$, and v be given. Write Y_0 for $A_s(y_0)$ and, for $\alpha > 0$, write B'_α for $B(y_0, \alpha s)$. Let K_1 and K_2 be some positive constants that are independent of y_0, s, and X, and will be chosen later. Pick $h_{K_2} \in C^\infty(\mathbb{R}^n)$ such that $h_{K_2} \equiv 1$ on $\mathbb{R}^n \setminus B'_{K_2}$, $0 \leq h_{K_2} \leq 1$ everywhere, and $h_{K_2} \equiv 0$ on $B_{K_2/2}$. Define u_{K_2} as the solution of $Lu_{K_2} = 0$ in Ω with the Dirichlet condition h_{K_2} on Γ, that will serve as a smooth substitute for $X \to \omega^X(\Gamma \setminus B'_{K_2})$. Define a function $f_{y_0,s}$ on $\Omega \setminus \{Y_0\}$ by

$$f_{y_0,s}(X) = s^{n-2} g(X, Y_0) - K_1 u_{K_2}. \tag{11.16}$$

When $|X - Y_0| \geq s/8$, the term $s^{n-2}g(X,Y_0)$ is uniformly bounded: this fact can be found in [**HoK**] (for $n \geq 3$) and [**DK**] (for $n = 2$). In addition, due to the non-degeneracy of the harmonic measure (same argument as for (11.10), similar to [**Ken**, Lemma 1.3.2]), there exists $C > 0$ (independent of $K_2 > 0$) such that $\omega^X(\Gamma \setminus B'_{K_2}) \geq C^{-1}$ for $X \in \Omega \setminus B'_{2K_2}$. Hence we can find $K_1 > 0$ such that for any choice of $K_2 > 0$, we have

$$f_{y_0,s}(X) \leq 0 \quad \text{for } X \in \Omega \setminus B'_{2K_2}. \tag{11.17}$$

For the sequel, we state an important result. There holds

$$C^{-1} s^{n-2} g(X, Y_0) \leq \omega^X(\Gamma \setminus B'_2) \leq C s^{n-2} g(X, Y_0) \quad \text{for } X \in \Omega \cap [B'_1 \setminus B(Y_0, s/8)], \tag{11.18}$$

where $C > 0$ depends only on n and the ellipticity constant of the matrix A. This result can be seen as an analogue of [**Ken**, Corollary 1.3.6]. It is proven in the higher codimension case in Lemma 11.11 below. The equivalence (11.18) can be seen as a weak version of the comparison principle, dealing only with harmonic measures and Green functions. It can be proven, like [**Ken**, Corollary 1.3.6], before the full comparison principle by using the specific properties of the Green functions and harmonic measures.

We want to take $K_2 > 0$ so large that

$$f_{y_0,s}(X) \geq \frac{1}{2} s^{n-2} g(X, Y_0) \quad \text{for } X \in \Omega \cap [B'_1 \setminus B(Y_0, s/8)]. \tag{11.19}$$

We build a smooth substitute u_4 for $\omega^X(\Gamma \setminus B'_2)$, namely the solution of $Lu_4 = 0$ in Ω with the Dirichlet condition $u_4 = h_4$ on Γ, where $h_4 \in C^\infty(\mathbb{R}^n)$, $h_4 \equiv 1$ on $\mathbb{R}^n \setminus B'_4$, $0 \leq h_4 \leq 1$ everywhere, and $h_4 \equiv 0$ on B'_2. Thanks to the Hölder continuity of solutions and the non-degeneracy of the harmonic measure, we have that for $X \in \Omega \cap [B'_{10} \setminus B'_5]$ and any $K_2 \geq 20$,

$$C^{-1} u_{K_2}(X) \leq (K_2)^{-\alpha} \leq C(K_2)^{-\alpha} u_4(X), \tag{11.20}$$

with constants $C, \alpha > 0$ independent of K_2, y_0, s or X. Since the functions u_{K_2} and u_4 are smooth enough, and $C^{-1}u_{K_2} = 0 \leq C(K_2)^{-\alpha} u_4(X)$ on $\Gamma \cap B'_{10}$, the maximum principle (Lemma 11.2) implies that

$$u_{K_2}(X) \leq C(K_2)^{-\alpha} u_4(X) \quad \text{for } X \in \Omega \cap B'_{10}. \tag{11.21}$$

We use (11.18) to get that for $K_2 \geq 20$,

$$K_1 u_{K_2}(X) \leq CK_1 (K_2)^{-\alpha} s^{n-2} g(X, Y_0) \quad \text{for } X \in \Omega \cap [B'_1 \setminus B(Y_0, s/8)]. \tag{11.22}$$

The inequality (11.19) can be now obtained by taking $K_2 \geq 20$ so that $CK_1(K_2)^{-\alpha} \leq \frac{1}{2}$. From (11.19) and (11.18), we deduce that

$$f_{y_0,s}(X) \geq C^{-1} \omega^X(\Gamma \setminus B'_2) \quad \text{for } X \in B'_1 \setminus B(Y_0, s/8), \tag{11.23}$$

where $C > 0$ depends only on n and the ellipticity constants of the matrix A.

Recall that our goal is to prove the claim (11.14), which will be established with $M = 4K_2$. Let v be a non-negative solution of $Lv = 0$ in $\Omega \cap B'_{4K_2}$. We can find $K_3 > 0$ (independent of y_0, s and X) such that

$$v(X) \geq K_3 v(Y_0) f_{y_0,s}(X) \quad \text{for } X \in B(Y_0, \frac{1}{4}s) \setminus B(Y_0, \frac{1}{8}s). \tag{11.24}$$

Indeed $f_{y_0,s}(X) \leq s^{n-2} g(X, Y_0) \leq C$ when $|X - Y_0| \geq s/8$, thanks to the pointwise bounds on the Green function (see [**HoK**], [**DK**]) and $v(X) \geq C^{-1} v(Y_0)$ when $|X - Y_0| \leq s/4$ because of the Harnack inequality. Also, thanks to (11.17),

$$v(X) \geq 0 \geq K_3 v(Y_0) f_{y_0,s}(X) \quad \text{for } X \in \Omega \cap [B'_{4K_2} \setminus B'_{2K_2}] \tag{11.25}$$

and it is easy to check that

$$v(y) \geq 0 \geq K_3 v(Y_0) f_{y_0,s}(y) \quad \text{for } y \in \Gamma \cap B'_{4K_2}. \tag{11.26}$$

We can apply our maximal principle, that is Lemma 11.2, with $E = B'_{4K_2} \setminus B(Y_0, \frac{1}{8}s)$ and $F = B'_{2K_2} \setminus B(Y_0, \frac{1}{4}s)$ and get that

$$v(X) \geq K_3 v(Y_0) f_{y_0,s}(X) \quad \text{for } X \in \Omega \cap [B'_{4K_2} \setminus B(Y_0, \frac{1}{8}s)]. \tag{11.27}$$

In particular, thanks to (11.23),

$$v(X) \geq C^{-1} v(Y_0) \omega^X(\Gamma \setminus B_2') \quad \text{for } X \in \Omega \cap [B_1' \setminus B(Y_0, \frac{1}{8}s)]. \tag{11.28}$$

Since both v and $X \to \omega^X(\Gamma \setminus B_2')$ are solutions in $\Omega \cap B_2'$, the Harnack inequality proves

$$v(X) \geq C^{-1} v(Y_0) \omega^X(\Gamma \setminus B_2') \quad \text{for } X \in \Omega \cap B_1'. \tag{11.29}$$

The claim (11.14) follows, which ends our alternative proof of Lemma 11.1.

11.2. The case of codimension higher than 1

We need first the following version of the maximum principle.

LEMMA 11.3. *Let $F \subset \mathbb{R}^n$ be a closed set and $E \subset \mathbb{R}^n$ an open set such that $F \subset E \subset \mathbb{R}^n$ and $\operatorname{dist}(F, \mathbb{R}^n \setminus E) > 0$. Let $u \in W_r(E)$ be a supersolution for L in $\Omega \cap E$ such that*

(i) $\int_E |\nabla u|^2 \, dm < +\infty$,
(ii) *$Tu \geq 0$ a.e. on $\Gamma \cap E$,*
(iii) *$u \geq 0$ a.e. in $(E \setminus F) \cap \Omega$.*

Then $u \geq 0$ a.e. in $E \cap \Omega$.

PROOF. The present proof is a slight variation of the proof of Lemma 9.2.

Set $v := \min\{u, 0\}$ in $E \cap \Omega$ and $v := 0$ in $\Omega \setminus E$. Note that $v \leq 0$. We want to use v as a test function. We claim that

$$v \text{ lies in } W_0 \text{ and is supported in } F. \tag{11.30}$$

Pick $\eta \in C_0^\infty(E)$ such that $\eta = 1$ in F and $\eta \geq 0$ everywhere, it is indeed possible because $\operatorname{dist}(F, \mathbb{R}^n \setminus E) > 0$. Since $u \in W_r(E)$, we have $\eta u \in W$, from which we deduce $\min\{0, \eta u\} \in W$ by Lemma 6.1. By (iii), $v = \min\{0, \eta u\}$ almost everywhere and hence $v \in W$.

Notice that $T(\eta u) \geq 0$ because of Assumption (ii) (and Lemma 8.1). Hence $v = \min\{\eta u, 0\} \in W_0$. And since (iii) also proves that v is supported in F, the claim (11.30) follows.

Since v is in W_0, Lemma 5.5 proves that v can be approached in W by a sequence of functions $(v_k)_{k\geq 1}$ in $C_0^\infty(\Omega)$ (i.e., that are compactly supported in Ω; see (5.24)). Note also that the construction used in Lemma 5.5 allows us, since $v \leq 0$ is supported in F, to take $v_k \leq 0$ and compactly supported in E. Definition 8.13 gives

$$\int_E \mathcal{A}\nabla u \cdot \nabla v_k \, dm = \int_\Omega \mathcal{A}\nabla u \cdot \nabla v_k \, dm \leq 0 \tag{11.31}$$

and since the map

$$\varphi \in W \to \int_E \mathcal{A}\nabla u \cdot \nabla \varphi \, dm \tag{11.32}$$

is bounded on W thanks to assumption (i) and (8.8), we deduce that

$$\int_E \mathcal{A}\nabla u \cdot \nabla v \, dm \leq 0. \tag{11.33}$$

Now Lemma 6.1 gives

$$\nabla v = \begin{cases} \nabla u & \text{if } u < 0 \\ 0 & \text{if } u \geq 0 \end{cases} \tag{11.34}$$

and so (11.33) becomes

$$\int_\Omega \mathcal{A}\nabla v \cdot \nabla v \, dm = \int_E \mathcal{A}\nabla u \cdot \nabla v \, dm \leq 0. \tag{11.35}$$

Together with the ellipticity condition (8.9), we obtain $\|v\|_W \leq 0$. Recall that $\|.\|_W$ is a norm on $W_0 \ni v$, hence $v = 0$ a.e. in Ω. We conclude from the definition of v that $u \geq 0$ a.e. in $E \cap \Omega$. □

Let us use the maximum principle above to prove the following result on the Green function.

LEMMA 11.4. *We have*

$$g(x,y) \leq C \min\{\delta(y), \delta(x)\}^{1-d} \quad \textit{for } x,y \in \Omega \textit{ such that } |x-y| \geq \delta(y)/4, \tag{11.36}$$

where the constant $C > 0$ depends only on d, n, C_0 and C_1.

REMARK 11.5. Lemma 11.4 is an improvement on the pointwise bounds (10.6) only when $d < 1$.

PROOF. Let $y \in \Omega$. Lemma 10.1 (v) gives

$$g(x,y) \leq K_1 \delta(y)^{1-d} \quad \text{for } x \in B(y, \delta(y)/4) \setminus B(y, \delta(y)/8) \tag{11.37}$$

for some $K_1 > 0$ that is independent of x and y. Define u on $\Omega \setminus \{y\}$ by $u(x) = K_1\delta(y)^{1-d} - g(x,y)$. Notice that u is a solution in $\Omega \setminus \overline{B(y, \delta(y)/4)}$, by (10.4). Also, thanks to Lemma 10.1 (i), the integral

$$\int_{\Omega \setminus B(y,\delta(y)/8)} |\nabla u|^2 \, dm = \int_{\Omega \setminus B(y,\delta(y)/8)} |\nabla g(.,y)|^2 \, dm$$

is finite, and $Tu = K_1\delta(y)^{1-d}$ is non-negative a.e. on Γ. In addition, due to (11.37), we have $u \geq 0$ on $B(y, \delta(y)/4) \setminus B(y, \delta(y)/4)$. Thus u satisfies all the assumption of Lemma 11.3 (the maximum principle), where we choose $E = \mathbb{R}^n \setminus \overline{B(y, \delta(y)/8)}$ and $F = \mathbb{R}^n \setminus B(y, \delta(y)/8)$, and which yields

$$g(x,y) \leq C\delta(y)^{1-d} \quad \text{for } x \in \Omega \setminus B(y, \delta(y)/8). \tag{11.38}$$

It remains to prove that

$$g(x,y) \leq C\delta(x)^{1-d} \quad \text{for } x, y \in \Omega \text{ such that } |x-y| \geq \delta(y)/4. \tag{11.39}$$

But Lemma 10.6 says that $g(x,y) = g_T(y,x)$, where g_T is the Green function associated to the operator $L_T = -\operatorname{div} A^T\nabla$. The above argument proves that

$$g(x,y) = g_T(y,x) \leq C\delta(x)^{1-d} \quad \text{for } x,y \in \Omega \text{ such that } |x-y| \geq \delta(x)/8, \tag{11.40}$$

which is (11.39) once we remark that $|x-y| \geq \delta(y)/4$ implies that $|x-y| \geq \delta(x)/8$. □

Let us prove the existence of "corkscrew points" in Ω.

LEMMA 11.6. *There exists $\epsilon > 0$, that depends only upon the dimensions d and n and the constant C_0, such that for $x_0 \in \Gamma$ and $r > 0$, there exists a point $A_r(x_0) \in \Omega$ such that*

(i) $|A_r(x_0) - x_0| \le r$,
(ii) $\delta(A_r(x_0)) \ge \epsilon r$.

In particular, $\delta(A_r(x_0)) \approx |A_r(x_0) - x_0| \approx r$.

In the sequel, for any $s > 0$ and $y \in \Gamma$, $A_s(y)$ will denote any point in Ω satisfying the conditions (i) and (ii) of Lemma 11.6.

PROOF. Let $x_0 \in \Gamma$ and $r > 0$ be given. Let $\epsilon \in (0, 1/8)$ be small, to be chosen soon. Let $z_1, \cdots z_N$ be a maximal collection of points of $B(x_0, (1-2\epsilon)r)$ that lie at mutual distances at least $4\epsilon r$. Set $B_i = B(z_i, \epsilon r)$; notice that the $2B_i = B(z_i, 2\epsilon r)$ are disjoint and contained in $B(x_0, r)$, and the $5B_i$ cover $B(x_0, (1-2\epsilon)r)$ (by maximality), so $\sum_i |5B_i| \ge |B(x_0, (1-2\epsilon)r)| \ge C^{-1}r^n$ and hence $N \ge C^{-1}\epsilon^{-n}$.

Suppose for a moment that every B_i meets Γ. Pick $y_i \in \Gamma \cap B_i$, notice that $B(y_i, \epsilon r) \subset 2B_i$, and then use the Ahlfors-regularity property (1.1) to prove that (11.41)

$$C_0^{-1}(\epsilon r)^d N \le \sum_{i=1}^N \mathcal{H}^d(\Gamma \cap B(y_i, \epsilon r)) \le \sum_{i=1}^N \mathcal{H}^d(\Gamma \cap 2B_i) \le \mathcal{H}^d(\Gamma \cap B(x_0, r)) \le C_0 r^d$$

because the $2B_i$ are disjoint and contained in $B(x_0, r)$. Thus $N \le C_0^2\epsilon^{-d}$, which makes our initial estimate on N impossible if we choose ϵ such that $\epsilon^{n-d} < C^{-1}C_0^{-2}$.

We pick ϵ like this, and by contraposition get that at least one B_i does not meet Γ. We choose $A_r(x_0) = z_i$, and notice that $\delta(x_i) \ge \epsilon r$ because $B_i \cap \Gamma = \emptyset$, and $|z_i - x_0| \le r$ by construction. The lemma follows. □

We also need the following slight improvement of Lemma 11.6.

LEMMA 11.7. *Let $M_1 \ge 1$ be given. There exists $M_2 > M_1$ (depending on d, n, C_0 and M_1) such that for any ball B of radius r and centered on Γ and any $x \in B \setminus \Gamma$ such that $\delta(x) \le \frac{r}{M_2}$, we can find $y \in B \setminus \Gamma$ such that*

(i) $\delta(y) \ge M_1\delta(x)$,
(ii) $|x - y| \le M_2\delta(x)$.

PROOF. The proof is almost the same. Let $M_1 \ge 1$ be given, and let $M_2 \ge 10M_1$ be large, to be chosen soon. Then let $B = B(x_0, r)$ and $x \in B$ be as in the statement. Set $B' = B(x_0, r - 2M_1\delta(x)) \cap B(x, (M_2 - 2M_1)\delta(x))$; notice that the two radii are larger than $M_2\delta(x)/2$, because $r \ge M_2\delta(x) \ge 10M_1\delta(x)$, so $|B'| \ge C^{-1}(M_2\delta(x))^n$ for a constant $C > 0$ independent of x.

Pick a maximal family (z_i), $1 \le i \le N$, of points of B' that lie at mutual distances at least $M_1\delta(x)$ from each other, and set $B_i = B(z_i, M_1\delta(x))$ for $1 \le i \le N$. The $5B_i$ cover B' by maximality, so $N \ge C^{-1}(M_1\delta(x))^{-n}|B'| \ge C^{-1}(M_2/M_1)^n$.

Suppose for a moment that every B_i meets Γ. Then pick $y_i \in B_i \cap \Gamma$ and use the Ahlfors regularity property (1.1) and the fact that the $2B_i$ contain the $B(y_i, M_1\delta(x))$ and are disjoint to prove that

$$\begin{aligned} C_0^{-1}(M_1\delta(x))^d N &\le \sum_{i=1}^N \mathcal{H}^d(\Gamma \cap B(y_i, M_1\delta(x))) \\ &\le \sum_{i=1}^N \mathcal{H}^d(\Gamma \cap 2B_i) \le \mathcal{H}^d(\Gamma \cap B(x, M_2\delta(x))) \le C_0(M_2\delta(x))^d. \qquad (11.42) \end{aligned}$$

That is, $M_1^d N \leq C_0^2 M_2^d$, and this contradicts our other bound for N if M_2/M_1 is large enough. We choose M_2 like this; then some B_i doesn't meet Γ, and we can take $y = z_i$. □

Before we prove the comparison theorem, we need a substitute for [**Ken**, Lemma 1.3.4].

LEMMA 11.8. *Let $x_0 \in \Gamma$ and $r > 0$ be given, and let $X_0 := A_r(x_0)$ be as in Lemma 11.6. Let $u \in W_r(B(x_0, 2r))$ be a non-negative, non identically zero, solution of $Lu = 0$ in $B(x_0, 2r) \cap \Omega$, such that $Tu \equiv 0$ on $B(x_0, 2r) \cap \Gamma$. Then*

$$u(X) \leq Cu(X_0) \qquad \text{for } X \in B(x_0, r), \tag{11.43}$$

where $C > 0$ depends only on d, n, C_0 and C_1.

PROOF. We follow the proof of [**KJ**, Lemma 4.4].

Let $x \in \Gamma$ and $s > 0$ such that $Tu \equiv 0$ on $B(x, s) \cap \Gamma$. Then the Hölder continuity of solutions given by Lemma 8.33 proves the existence of $\epsilon > 0$ (that depends only on d, n, C_0, C_1) such that

$$\sup_{B(x,\epsilon s)} u \leq \frac{1}{2} \sup_{B(x,s)} u. \tag{11.44}$$

Without loss of generality, we can choose $\epsilon < \frac{1}{2}$.

A rough idea of the proof of (11.43) is that $u(x)$ should not be near the maximum of u when x lies close to $B(x_0, r) \cap \Gamma$, because of (11.44). Then we are left with points x that lie far from the boundary, and we can use the Harnack inequality to control $u(x)$. The difficulty is that when $x \in B(x_0, r)$ lies close to Γ, $u(x)$ can be bounded by values of u inside the domain, and not by values of u near Γ but from the exterior of $B(x_0, r)$. We will prove this latter fact by contradiction: we show that if $\sup_{B(x_0,r)} u$ exceeds a certain bound, then we can construct a sequence of points $X_k \in B(x_0, \frac{3}{2}r)$ such that $\delta(X_k) \to 0$ and $u(X_k) \to +\infty$, and hence we contradict the Hölder continuity of solutions at the boundary.

Since $u(X) > 0$ somewhere, the Harnack inequality (Lemma 8.9), maybe applied a few times, yields $u(X_0) > 0$. We can rescale u and assume that $u(X_0) = 1$. We claim that there exists $M > 0$ such that for any integer $N \geq 1$ and $Y \in B(x_0, \frac{3}{2}r)$,

$$\delta(Y) \geq \epsilon^N r \Longrightarrow u(Y) \leq M^N, \tag{11.45}$$

where ϵ comes from (11.44) and the constant M depends only upon d, n, C_0, C_1. We will prove this by induction. The base case (and in fact we will manage to start directly from some large integer N_0) is given by the following. Let $M_2 > 0$ be the value given by Lemma 11.7 when $M_1 := \frac{1}{\epsilon}$. Let $N_0 \geq 1$ be the smallest integer such that $M_2 \leq \epsilon^{-N_0}$. We want to show the existence of $M_3 \geq 1$ such that

$$u(Y) \leq M_3 \qquad \text{for every } Y \in B(x_0, \frac{3}{2}r) \text{ such that } \delta(Y) \geq \epsilon^{N_0} r. \tag{11.46}$$

Indeed, if $Y \in B(x_0, \frac{3}{2}r)$ satisfies $\delta(Y) \geq \epsilon^{N_0} r$, Lemma 2.2 and the fact that $|x_0 - X_0| \approx r$ (by Lemma 11.6) imply the existence of a Harnack chain linking Y to X_0. More precisely, we can find balls $B_1, \dots, B_h$ with a same radius, such that $Y \in B_1$, $X_0 \in B_h$, $3B_i \subset B(x_0, 2r) \setminus \Gamma$ for $i \in \{1, \dots, h\}$, and $B_i \cap B_{i+1} \neq \emptyset$ for $i \in \{1, \dots, h-1\}$, and in addition h is bounded independently of x_0, r and Y.

Together with the Harnack inequality (Lemma 8.9), we obtain (11.46). This proves (11.45) for $N = N_0$, but also directly for $1 \leq N \leq N_0$, if we choose $M \geq M_3$.

For any point $Y \in B(x_0, \frac{3}{2}r)$ such that $\delta(Y) \leq \epsilon^{N_0} r \leq \frac{r}{M_2}$, Lemma 11.7 (and our choice of M_2) gives the existence of $Z \in B(x_0, \frac{3}{2}r) \cap B(Y, M_2\delta(Y))$ such that $\delta(Z) \geq M_1\delta(Y)$. Since $Z \in B(Y, M_2\delta(Y))$ and $\delta(Z) > \delta(Y) > 0$, Lemma 2.2 implies the existence of a Harnack chain whose length is bounded by a constant depending on d, n, C_0 (and M_2 - but M_2 depends only on the three first parameters) and together with the Harnack inequality (Lemma 8.9), we obtain the existence of $M_4 \geq 1$ (that depends only on d, n, C_0 and C_1) such that $u(Y) \leq M_4 u(Z)$. So we just proved that

(11.47)
$$\text{for any } Y \in B(x_0, \tfrac{3}{2}r) \text{ such that } \delta(Y) \leq \epsilon^{N_0} r, \\ \text{there exists } Z \in B(x_0, \tfrac{3}{2}r) \text{ such that } \delta(Z) \geq M_1\delta(Y) \text{ and } u(Y) \leq M_4 u(Z).$$

We turn to the main induction step. Set $M = \max\{M_3, M_4\} \geq 1$ and let $N \geq N_0$ be given. Assume, by induction hypothesis, that for any $Z \in B(x_0, \frac{3}{2}r)$ satisfying $\delta(Z) \geq \epsilon^N r$, we have $u(Z) \leq M^N$. Let $Y \in B(x_0, \frac{3}{2}r)$ be such that $\delta(Y) \geq \epsilon^{N+1} r$. The assertion (11.47) yields the existence of $Z \in B(x_0, \frac{3}{2}r)$ such that $\delta(Z) \geq M_1\delta(Y) = \epsilon^{-1}\delta(Y) \geq \epsilon^N r$ and $u(Y) \leq M_4 u(Z) \leq M u(Z)$. By the induction hypothesis, $u(Y) \leq M^{N+1}$. This completes our induction step, and the proof of (11.45) for every $N \geq 1$.

Choose an integer i such that $2^i \geq M$, where M is the constant of (11.45) that we just found, and then set $M' = M^{i+3}$. We want to prove by contradiction that

$$u(X) \leq M' u(X_0) = M' \quad \text{for every } X \in B(x_0, r). \tag{11.48}$$

So we assume that

$$\text{there exists } X_1 \in B(x_0, r) \text{ such that } u(X_1) > M' \tag{11.49}$$

and we want to prove by induction that for every integer $k \geq 1$,

(11.50)
$$\text{there exists } X_k \in B(x_0, \frac{3}{2}r) \text{ such that } u(X_k) > M^{i+2+k} \text{ and } |X_k - x_0| \leq \frac{3}{2}r - 2^{-k}r.$$

The base step of the induction is given by (11.49) and we want to do the induction step. Let $k \geq 1$ be given and assume that (11.50) holds. From the contraposition of (11.45), we deduce that $\delta(X_k) < \epsilon^{i+2+k} r$. Choose $x_k \in \Gamma$ such that $|X_k - x_k| = \delta(X_k) < \epsilon^{i+2+k} r$. By the induction hypothesis,

$$|x_k - x_0| \leq |x_k - X_k| + |X_k - x_0| \leq \frac{3r}{2} - 2^{-k}r + \epsilon^{i+2+k} r \tag{11.51}$$

and, since $\epsilon \leq \frac{1}{2}$,

$$|x_k - x_0| \leq \frac{3r}{2} - 2^{-k}r + 2^{-2-k}r. \tag{11.52}$$

Now, due to (11.44), we can find $X_{k+1} \in B(x_k, \epsilon^{2+k} r)$ such that

$$u(X_{k+1}) \geq 2^i \sup_{X \in B(x_k, \epsilon^{i+2+k} r)} u(X) \geq 2^i u(X_k) \geq M^{i+2+(k+1)}. \tag{11.53}$$

The induction step will be complete if we can prove that $|X_{k+1}-x_0| \leq \frac{3}{2}r-2^{-(k+1)}r$. Indeed,

$$
\begin{aligned}
|X_{k+1}-x_0| &\leq |X_{k+1}-x_k| + |x_k - x_0| \leq \epsilon^{2+k}r + \frac{3r}{2} - 2^{-k}r + 2^{-2-k}r \\
&\leq \frac{3r}{2} - 2^{-k}r + 2^{-1-k}r = \frac{3}{2}r - 2^{-k-1}r
\end{aligned}
\tag{11.54}
$$

by (11.52) and because $\epsilon \leq \frac{1}{2}$.

Let us sum up. We assumed the existence of $X_1 \in B(x_0,r)$ such that $u(X_1) > M'$ and we end up with (11.50), that is a sequence X_k of values in $B(x_0,\frac{3}{2}r)$ such that $u(X_k)$ increases to $+\infty$. Up to a subsequence, we can thus find a point in $\overline{B(x_0,\frac{3}{2}r)}$ where u is not continuous, which contradicts Lemma 8.16. Hence $u(X) \leq M' = M'u(X_0)$ for $X \in B(x_0,r)$. Lemma 11.8 follows. $\square$

LEMMA 11.9. *Let $x_0 \in \Gamma$ and $r > 0$ be given, and set $X_0 := A_r(x_0)$ as in Lemma 11.6. Then for all $X \in \Omega \setminus B(X_0,\delta(X_0)/4)$,*

$$r^{d-1}g(X,X_0) \leq C\omega^X(B(x_0,r)\cap\Gamma) \tag{11.55}$$

and

$$r^{d-1}g(X,X_0) \leq C\omega^X(\Gamma\setminus B(x_0,2r)), \tag{11.56}$$

where $C>0$ depends only on d, n, C_0 and C_1.

PROOF. We prove (11.55) first. Let $h \in C_0^\infty(B(x_0,r))$ satisfy $h \equiv 1$ on $B(x_0,r/2)$ and $0 \leq h \leq 1$. Define then $u \in W$ as the solution of $Lu = 0$ with data Th given by Lemma 9.1. Set $v(X) = 1 - u(X) \in W$ and observe that $0 \leq v \leq 1$ and $Tv = 0$ on $B(x_0,r/2)\cap\Gamma$.

By Lemma 8.16, we can find $\epsilon > 0$ (that depends only on d, n, C_0, C_1) such that $v(A_{\epsilon r}(x_0)) \leq \frac{1}{2}$, i.e. $u(A_{\epsilon r}(x_0)) \geq \frac{1}{2}$. The existence of Harnack chains (Lemma 2.2) and the Harnack inequality (Lemma 8.9) give

$$C^{-1} \leq u(X) \quad \text{for } X \in B(X_0,\delta(X_0)/2). \tag{11.57}$$

By Lemma 10.1 (v), $g(X,X_0) \leq C|X-X_0|^{1-d}$ for $X \in \Omega\setminus B(X_0,\delta(X_0)/4)$. Since $\delta(X_0) \approx r$ by construction of X_0,

$$r^{d-1}g(X,X_0) \leq C \quad \text{for } X \in B(X_0,\delta(X_0)/2)\setminus B(X_0,\delta(X_0)/4). \tag{11.58}$$

The combination of (11.57) and (11.58) yields the existence of $K_1 > 0$ (depending only on n, d, C_0 and C_1) such that

$$r^{d-1}g(X,X_0) \leq K_1 u(X) \quad \text{for } X \in B(X_0,\delta(X_0)/2)\setminus B(X_0,\delta(X_0)/4). \tag{11.59}$$

We claim that $K_1u(X) - r^{d-1}g(X,X_0)$ satisfies the assumptions of Lemma 11.3, with $E = \mathbb{R}^n \setminus \overline{B(X_0,\delta(X_0)/4)}$ and $F = \mathbb{R}^n\setminus B(X_0,\delta(X_0)/2)$. Indeed Assumption (i) of Lemma 11.3 is satisfied because $u \in W$ and by Lemma 10.1 (i). Assumption (ii) of Lemma 11.3 holds because $Tu = h \geq 0$ by construction and also $Tg(.,X_0) = 0$ thanks to Lemma 10.1 (i). Assumption (iii) of Lemma 11.3 is given by (11.59). The lemma yields

$$r^{d-1}g(X,X_0) \leq K_1u(X) \quad \text{for } X \in \Omega\setminus B(X_0,\delta(X_0)/4). \tag{11.60}$$

By the positivity of the harmonic measure, $u(X) \leq \omega^X(B(x_0,r)\cap\Gamma)$ for $X \in \Omega$; (11.55) follows.

Let us turn to the proof of (11.56). We want to find two points $x_1, x_2 \in \Gamma \cap [B(x_0, Kr) \setminus B(x_0, 4r)]$, where the constant $K \geq 10$ depends only on C_0 and d, such that $X_1 := A_r(x_1)$ and $X_2 := A_r(x_2)$ satisfy

$$B(X_1, \delta(X_1)/4) \cap B(X_2, \delta(X_2)/4) = \emptyset. \tag{11.61}$$

To get such points, we use the fact that Γ is Ahlfors regular to find $M \geq 3$ (that depends only on C_0 and d) such that $\Gamma_1 := \Gamma \cap [B(x_0, 2Mr) \setminus B(x_0, 6r)] \neq \emptyset$ and $\Gamma_2 := \Gamma \cap [B(x_0, 2M^2 r) \setminus B(x_0, 6Mr)] \neq \emptyset$. Any choice of points $x_1 \in \Gamma_1$ and $x_2 \in \Gamma_2$ verifies (11.61).

Let $X \in \Omega \setminus B(X_0, \delta(X_0)/4)$. Thanks to (11.61), there exists $i \in \{1, 2\}$ such that $X \notin B(X_i, \delta(X_i)/4)$. The existence of Harnack chains (Lemma 2.2), the Harnack inequality (Lemma 8.9), and the fact that $Y \to g(X, Y)$ is a solution of $L_T u := -\operatorname{div} A^T \nabla u = 0$ in $\Omega \setminus \{X\}$ (Lemma 10.1 and Lemma 10.6) yield

$$r^{d-1} g(X, X_0) \leq C r^{d-1} g(X, X_i). \tag{11.62}$$

By (11.55) and the positivity of the harmonic measure,

$$r^{d-1} g(X, X_0) \leq C r^{d-1} g(X, X_i) \leq C w^X(B(x_i, r) \cap \Gamma) \leq C w^X(\Gamma \setminus B(x_0, r)). \tag{11.63}$$

The lemma follows. □

We turn now to the non-degeneracy of the harmonic measure.

LEMMA 11.10. *Let $\alpha > 1$, $x_0 \in \Gamma$, and $r > 0$ be given, and let $X_0 := A_r(x_0) \in \Omega$ be as in Lemma 11.6. Then*

$$\omega^X(B(x_0, r) \cap \Gamma) \geq C_\alpha^{-1} \quad \textit{for } X \in B(x_0, r/\alpha), \tag{11.64}$$

$$\omega^X(B(x_0, r) \cap \Gamma) \geq C_\alpha^{-1} \quad \textit{for } X \in B(X_0, \delta(X_0)/\alpha), \tag{11.65}$$

$$\omega^X(\Gamma \setminus B(x_0, r)) \geq C_\alpha^{-1} \quad \textit{for } X \in \Omega \setminus B(x_0, \alpha r), \tag{11.66}$$

and

$$\omega^X(\Gamma \setminus B(x_0, r)) \geq C_\alpha^{-1} \quad \textit{for } X \in B(X_0, \delta(X_0)/\alpha), \tag{11.67}$$

where $C_\alpha > 0$ depends only upon d, n, C_0, C_1 and α.

PROOF. Let us first prove (11.64). Set $u(X) = 1 - \omega^X(B(x_0, r) \cap \Gamma)$. By Lemma 9.7, u lies in $W_r(B(x_0, r))$, is a solution of $Lu = 0$ in $\Omega \cap B(x_0, r)$, and has a vanishing trace on $\Gamma \cap B(x_0, r)$. So the Hölder continuity of solutions at the boundary (Lemma 8.16) gives the existence of an $\epsilon > 0$, that depends only on d, n, C_0, C_1 and α, such that $u(X) \leq \frac{1}{2}$ for every $X \in B(x_0, \frac{1}{2}[1 + \frac{1}{\alpha}]r)$ such that $\delta(X) \leq \epsilon r$. Thus $v(X) := \omega^X(B(x_0, r) \geq \frac{1}{2}$ for $X \in B(x_0, \frac{1}{2}[1 + \frac{1}{\alpha}]r)$ such that $\delta(X) \leq \epsilon r$. We now deduce (11.64) from the existence of Harnack chains (Lemma 2.2) and the Harnack inequality (Lemma 8.9).

The assertion (11.65) follows from (11.64). Indeed, (11.64) implies that

$$\omega^{A_{r/2}(x_0)}(B(x_0, r) \cap \Gamma) \geq C^{-1}.$$

The existence of Harnack chains (Lemma 2.2) and the Harnack inequality (Lemma 8.9) allow us to conclude. Finally (11.66) and (11.67) can be proved as above, and we leave the details to the reader. □

LEMMA 11.11. *Let $x_0 \in \Gamma$ and $r > 0$ be given, and set $X_0 = A_r(x_0)$. Then*

(11.68)
$$C^{-1}r^{d-1}g(X, X_0) \leq \omega^X(B(x_0, r) \cap \Gamma) \leq Cr^{d-1}g(X, X_0) \quad \text{for } X \in \Omega \setminus B(x_0, 2r),$$

and

(11.69)
$$C^{-1}r^{d-1}g(X, X_0) \leq \omega^X(\Gamma \setminus B(x_0, 2r)) \leq Cr^{d-1}g(X, X_0)$$
$$\text{for } X \in B(x_0, r) \setminus B(X_0, \delta(X_0)/4),$$

where $C > 0$ depends only upon d, n, C_0 and C_1.

PROOF. The lower bounds are a consequence of Lemma 11.9; the one in (11.68) also requires to notice that $\delta(X_0) \leq r$ and thus $B(X_0, \delta(X_0)/4) \subset B(x_0, 2r)$.

It remains to check the upper bounds. But we first prove an intermediate result. We claim that for $\phi \in C^\infty(\mathbb{R}^n) \cap W$ and $X \notin \operatorname{supp} \phi$,

$$u_\phi(X) = -\int_\Omega A\nabla\phi(Y) \cdot \nabla_y g(X, Y) dY, \tag{11.70}$$

where $u_\phi \in W$ is the solution of $Lu_\phi = 0$, with the Dirichlet condition $Tu_\phi = T\phi$ on Γ, given by Lemma 9.1. Indeed, recall that by (8.8) and (8.9) the map

$$u, v \in W_0 \to \int_\Omega A\nabla u \cdot \nabla v\, dY = \int_\Omega \mathcal{A}\nabla u \cdot \nabla v\, dm \tag{11.71}$$

is bounded and coercive on W_0 and the map

$$\varphi \in W_0 \to \int_\Omega A\nabla\phi \cdot \nabla\varphi\, dY = \int_\Omega \mathcal{A}\nabla\phi \cdot \nabla\varphi\, dm \tag{11.72}$$

is bounded on W_0. So the Lax-Milgram theorem yields the existence of $\mathfrak{v} \in W_0$ such that

$$\int_\Omega A\nabla\phi \cdot \nabla\varphi\, dY = \int_\Omega A\nabla\mathfrak{v} \cdot \nabla\varphi\, dY \qquad \forall\varphi \in W_0. \tag{11.73}$$

Let $s > 0$ such that $B(X, 2s) \cap (\operatorname{supp}\phi \cup \Gamma) = \emptyset$. For any $\rho > 0$ we define, as we did in (10.16), the function $\mathfrak{g}_T^\rho = g_T^\rho(., X)$ on Ω as the only function in W_0 such that

$$\int_\Omega A\nabla\varphi \cdot \nabla\mathfrak{g}_T^\rho\, dY = \int_\Omega A^T\nabla\mathfrak{g}_T^\rho \cdot \nabla\varphi\, dY = \fint_{B(X,\rho)} \varphi\, dY \qquad \forall\varphi \in W_0. \tag{11.74}$$

We take $\varphi = \mathfrak{g}_T^\rho$ in (11.73) to get

$$\int_\Omega A\nabla\phi \cdot \nabla\mathfrak{g}_T^\rho\, dY = \int_\Omega A\nabla\mathfrak{v} \cdot \nabla\mathfrak{g}_T^\rho\, dY = \fint_{B(X,\rho)} \mathfrak{v}\, dY. \tag{11.75}$$

We aim to take the limit as $\rho \to 0$ in (11.75). Since $\mathfrak{v}$ satisfies

$$\int_\Omega A\nabla\mathfrak{v} \cdot \nabla\varphi\, dY = \int_\Omega A\nabla\phi \cdot \nabla\varphi\, dY = 0 \qquad \forall\varphi \in C_0^\infty(B(X, 2s)), \tag{11.76}$$

$\mathfrak{v}$ is a solution of $L\mathfrak{v} = 0$ on $B(X, 2s)$ and thus Lemma 8.8 proves that $\mathfrak{v}$ is continuous at X. As a consequence,

$$\lim_{\rho\to 0} \fint_{B(X,\rho)} \mathfrak{v}\, dY = \mathfrak{v}(X). \tag{11.77}$$

Recall that the $\mathfrak{g}_T^\rho$, $\rho > 0$, are the same functions as in in the proof of Lemma 10.1, but for the transpose matrix A^T. Let $\alpha \in C_0^\infty(B(x, 2s))$ be such that $\alpha \equiv 1$ on

$B(x,s)$. By (10.78) and Lemma 10.6, there exists a sequence (ρ_η) tending to 0, such that $(1-\alpha)\mathfrak{g}_T^{\rho_\eta}$ converges weakly to $(1-\alpha)g_T(.,X) = (1-\alpha)g(X,.)$ in W_0. As a consequence,

$$\begin{aligned}\lim_{\eta\to+\infty}\int_\Omega A\nabla\phi\cdot\nabla\mathfrak{g}_T^{\rho_\eta}\,dY &= \lim_{\eta\to+\infty}\int_\Omega A\nabla\phi\cdot\nabla[(1-\alpha)\mathfrak{g}_T^{\rho_\eta}]\,dY\\ &= \int_\Omega A\nabla\phi(Y)\cdot\nabla_y[(1-\alpha)g(X,Y)]dY\\ &= \int_\Omega A\nabla\phi(Y)\cdot\nabla_y g(X,Y)dY. \end{aligned} \tag{11.78}$$

The combination of (11.75), (11.77) and (11.78) yields

$$\int_\Omega A\nabla\phi(Y)\cdot\nabla_y g(X,Y)dY = \mathfrak{v}(X). \tag{11.79}$$

Since $\mathfrak{v}\in W_0$ satisfies (11.73), the function $u_\phi = \phi - \mathfrak{v}$ lies in W and is a solution of $Lu_\phi = 0$ with the Dirichlet condition $Tu_\phi = T\phi$. Hence

$$\int_\Omega A\nabla\phi(Y)\cdot\nabla_y g(X,Y)dY = \mathfrak{v}(X) = \phi(X) - u_\phi(X) = -u_\phi(X), \tag{11.80}$$

by (11.79) and because $X\notin \operatorname{supp}\phi$. The claim (11.70) follows.

We turn to the proof of the upper bound in (11.68), that is,

$$\omega^X(B(x_0,r)\cap\Gamma) \leq Cr^{d-1}g(X,X_0) \quad \text{for } X\in\Omega\setminus B(x_0,2r). \tag{11.81}$$

Let $X\in\Omega\setminus B(x_0,2r)$ be given, and choose $\phi\in C_0^\infty(\mathbb{R}^n)$ such that $0\leq\phi\leq 1$, $\phi\equiv 1$ on $B(x_0,r)$, $\phi\equiv 0$ on $\mathbb{R}^n\setminus B(x_0,\frac{5}{4}r)$, and $|\nabla\phi|\leq\frac{10}{r}$. We get that

$$u_\phi(X) \leq \frac{C}{r}\int_{B(x_0,\frac{5}{4}r)} |\nabla_y g(X,Y)|dm(Y) \tag{11.82}$$

by (11.70) and (8.8), and since $\omega^X(B(x_0,r)\cap\Gamma)\leq u_\phi(X)$ by the positivity of the harmonic measure,

$$\begin{aligned}\omega^X(B(x_0,r)\cap\Gamma) &\leq \frac{C}{r}\int_{B(x_0,\frac{5}{4}r)} |\nabla_y g(X,Y)|dm(Y).\\ &\leq \frac{C}{r}r^{\frac{d+1}{2}}\left(\int_{B(x_0,\frac{5}{4}r)} |\nabla_y g(X,Y)|^2dm(Y)\right)^{\frac{1}{2}}\end{aligned} \tag{11.83}$$

by Cauchy-Schwarz' inequality and Lemma 2.3. Since $X\in\Omega\setminus B(x_0,2r)$, Lemma 10.6 and Lemma 10.1 (iii) say that the function $Y\to g(X,Y)$ is a solution of $L_Tu := -\operatorname{div} A^T\nabla u$ on $B(x_0,2r)$, with a vanishing trace on $\Gamma\cap B(x_0,2r)$. So the Caccioppoli inequality at the boundary (see Lemma 8.11) applies and yields

$$\omega^X(B(x_0,r)\cap\Gamma) \leq \frac{C}{r^2}r^{\frac{d+1}{2}}\left(\int_{B(x_0,\frac{3}{2}r)} |g(X,Y)|^2dm(Y)\right)^{\frac{1}{2}}. \tag{11.84}$$

Then by Lemma 11.8,

$$\omega^X(B(x_0,r)\cap\Gamma) \leq \frac{C}{r^2}r^{d+1}g(X,X_0) = Cr^{d-1}g(X,X_0); \tag{11.85}$$

the bound (11.81) follows.

It remains to prove the upper bound in (11.69), i.e., that

$$\omega^X(\Gamma \setminus B(x_0,2r)) \leq Cr^{d-1}g(X,X_0) \quad \text{for } X \in B(x_0,r)\setminus B(X_0,\delta(X_0)/4). \tag{11.86}$$

The proof will be similar to the upper bound in (11.68) once we choose an appropriate function ϕ in (11.70). Let us do this rapidly. Let $X \in B(x_0,r)\setminus B(X_0,\delta(X_0)/4)$ be given and take $\phi \in C^\infty(\mathbb{R}^n)$ such that $0 \leq \phi \leq 1$, $\phi \equiv 1$ on $\mathbb{R}^n \setminus B(x_0,\frac{8}{5}r)$, $\phi \equiv 0$ on $B(x_0,\frac{7}{5}r)$ and $|\nabla\phi| \leq \frac{10}{r}$. Notice that $X \notin \operatorname{supp}(\phi)$, so (11.70) applies and yields

$$u_\phi(X) \leq \frac{C}{r}\int_{B(x_0,\frac{8}{5}r)\setminus B(x_0,\frac{7}{5}r)} |\nabla_y g(X,Y)|dm(Y). \tag{11.87}$$

By the positivity of the harmonic measure, $\omega^X(\Gamma\setminus B(x_0,2r)) \leq u_\phi(X)$. We use the Cauchy-Schwarz and Caccioppoli inequalities (see Lemma 8.11), as above, and get that

$$\begin{aligned}
&\omega^X(\Gamma\setminus B(x_0,2r))\\
&\leq \frac{C}{r}m(B(x_0,\frac{8}{5}r))\left(\frac{1}{m(B(x_0,\frac{8}{5}r))}\int_{B(x_0,\frac{8}{5}r)\setminus B(x_0,\frac{7}{5}r)}|\nabla_y g(X,Y)|^2dm(Y)\right)^{\frac12}\\
&\leq \frac{C}{r^2}r^{d+1}\left(\frac{1}{m(B(x_0,\frac{8}{5}r))}\int_{B(x_0,\frac{9}{5}r)\setminus B(x_0,\frac{6}{5}r)}|g(X,Y)|^2dm(Y)\right)^{\frac12}.
\end{aligned} \tag{11.88}$$

We claim that

$$g(X,Y) \leq Cg(X,X_0) \qquad \forall Y \in B(x_0,\frac{9}{5}r)\setminus B(x_0,\frac{6}{5}r) \tag{11.89}$$

where $C>0$ depends only on d, n, C_0 and C_1. Two cases may happen. If $\delta(Y) \geq \frac{r}{20}$, (11.89) is only a consequence of the existence of Harnack chains (Lemma 2.2) and the Harnack inequality (Lemma 8.9). Otherwise, if $\delta(Y) < \frac{r}{20}$ then Lemma 11.8 says that $g(X,Y) \leq Cg(X,X_Y)$ for some point $X_Y \in B(x_0,\frac{9}{5}r)\setminus B(x_0,\frac{6}{5}r)$ that lies at distance at least ϵr from Γ. Here ϵ comes from Lemma 11.6 and thus depends only on d, n and C_0. Together with the existence of Harnack chains (Lemma 2.2) and the Harnack inequality (Lemma 8.9), we find that $g(X,X_Y)$, or $g(X,Y)$, is bounded by $Cg(X,X_0)$.

We use (11.89) in the right hand side of (11.88) to get that

$$\omega^X(\Gamma\setminus B(x_0,2r)) \leq Cr^{d-1}\frac{m(B(x_0,\frac{9}{5}r))}{m(B(x_0,\frac{8}{5}r))}g(X,X_0) \leq Cr^{d-1}g(X,X_0), \tag{11.90}$$

by the doubling property of the measure m. The second and last assertion of the lemma follows. □

LEMMA 11.12 (Doubling volume property for the harmonic measure). *For $x_0 \in \Gamma$ and $r>0$, we have*

$$\omega^X(B(x_0,2r)\cap\Gamma) \leq C\omega^X(B(x_0,r)\cap\Gamma) \quad \textit{for } X \in \Omega\setminus B(x_0,4r) \tag{11.91}$$

and

$$\omega^X(\Gamma\setminus B(x_0,r)) \leq C\omega^X(\Gamma\setminus B(x_0,2r)) \quad \textit{for } X \in B(x_0,r/2), \tag{11.92}$$

where $C>0$ depends only on n, d, C_0 and C_1.

PROOF. Let us prove (11.91) first. Lemma 11.11 says that for $X \in \Omega\backslash B(x_0, 4r)$,

$$\omega^X(B(x_0, 2r) \cap \Gamma) \approx r^{d-1} g(X, A_{2r}(x_0)) \tag{11.93}$$

and

$$\omega^X(B(x_0, r) \cap \Gamma) \approx r^{d-1} g(X, A_r(x_0)), \tag{11.94}$$

where $A_{2r}(x_0)$ and $A_r(x_0)$ are the points of Ω given by Lemma 11.6. The bound (11.91) will be thus proven if we can show that

$$g(X, A_{2r}(x_0)) \approx g(X, A_r(x_0)) \quad \text{ for } X \in \Omega \setminus B(x_0, 4r). \tag{11.95}$$

Yet, since $Y \to g(X, Y)$ belongs to $W_r(\Omega \setminus \{X\})$ and is a solution of $L_T u := -\operatorname{div} A^T \nabla u = 0$ in $\Omega \setminus \{X\}$ (see Lemma 10.1 and Lemma 10.6), the equivalence in (11.95) is an easy consequence of the properties of $A_r(x_0)$ (Lemma 11.6), the existence of Harnack chains (Lemma 2.2) and the Harnack inequality (Lemma 8.9).

We turn to the proof of (11.92). Set $X_1 := A_r(x_0)$ and $X_{\frac{1}{2}} := A_{r/2}(x_0)$. Call Ξ the set of points $X \in B(x_0, r/2)$ such that $|X - X_1| \geq \frac{1}{4}\delta(X_1)$ and $|X - X_{\frac{1}{2}}| \geq \frac{1}{4}\delta(X_{\frac{1}{2}})$, and first consider $X \in \Xi$. By Lemma 11.11 again,

$$\omega^X(\Gamma \setminus B(x_0, 2r)) \approx r^{d-1} g(X, X_1) \tag{11.96}$$

and

$$\omega^X(\Gamma \setminus B(x_0, r)) \approx r^{d-1} g(X, X_{\frac{1}{2}}). \tag{11.97}$$

Since $\delta(X_1) \approx \delta(X_{\frac{1}{2}}) \approx r$ and $Y \to g(X, Y)$ is a solution of $L_T u = -\operatorname{div} A^T \nabla u = 0$, the existence of Harnack chains (Lemma 2.2) and the Harnack inequality (Lemma 8.9) give $g(X, X_1) \approx g(X, X_{\frac{1}{2}})$ for $X \in \Xi$. Hence

$$\omega^X(\Gamma \setminus B(x_0, 2r)) \approx \omega^X(\Gamma \setminus B(x_0, r)), \tag{11.98}$$

with constants that do not depend on X, x_0, or r. The equivalence in (11.98) also holds for all $X \in B(x_0, r/2)$, and not only for $X \in \Xi$, by Harnack's inequality (Lemma 8.9). This proves (11.92). □

REMARK 11.13. The following results also hold for every $\alpha > 1$. For $x_0 \in \Gamma$ and $r > 0$,

$$\omega^X(B(x_0, 2r) \cap \Gamma) \leq C_\alpha \omega^X(B(x_0, r) \cap \Gamma) \quad \text{ for } X \in \Omega \setminus B(x_0, 2\alpha r), \tag{11.99}$$

and

$$\omega^X(\Gamma \setminus B(x_0, r)) \leq C_\alpha \omega^X(\Gamma \setminus B(x_0, 2r)) \quad \text{ for } X \in B(x_0, r/\alpha), \tag{11.100}$$

where $C_\alpha > 0$ depends only on n, d, C_0, C_1 and α.

This can be deduced from Lemma 11.12 - that corresponds to the case $\alpha = 2$ - by applying it to smaller balls.

Let us prove for instance (11.100). Let $X \in B(x_0, r/\alpha)$ be given. We only need to prove (11.100) when $\delta(X) < \frac{r}{4}(1 - \frac{1}{\alpha})$, because as soon as we do this, the other case when $\delta(X) \geq \frac{r}{4}(1 - \frac{1}{\alpha})$ follows, by Harnack's inequality (Lemma 8.9).

Let $x \in \Gamma$ such that $|x - X| = \delta(X)$; then set $r_k = 2^{k-1} r[1 - \frac{1}{\alpha}]$ and $B_k = B(x, r_k)$ for $k \in \mathbb{Z}$. We wish to apply the doubling property (11.92) and get that

$$\omega^X(\Gamma \setminus B_k) \leq C \omega^X(\Gamma \setminus B_{k+1}), \tag{11.101}$$

and we can do this as long as $X \in B_{k-1}$. With our extra assumption that $|x-X| = \delta(X) < \frac{r}{4}(1-\frac{1}{\alpha})$, this is possible for all $k \geq 0$. Notice that

$$|x-x_0| \leq \delta(X)+|X-x_0| \leq \frac{r}{4}(1-\frac{1}{\alpha})+\frac{r}{\alpha} \leq \frac{r}{2}(1-\frac{1}{\alpha})+\frac{r}{\alpha} = \frac{r}{2}[1+\frac{1}{\alpha}] \tag{11.102}$$

and then $|x-x_0|+r_0 \leq \frac{r}{2}[1+\frac{1}{\alpha}]+\frac{r}{2}[1-\frac{1}{\alpha}] = r$, so $B_0 = B(x, r_0) \subset B(x_0, r)$ and, by the monotonicity of the harmonic measure,

$$\omega^X(\Gamma \setminus B(x_0, r)) \leq \omega^X(\Gamma \setminus B_0). \tag{11.103}$$

Let k be the smallest integer such that $2^{k-1}(1-\frac{1}{\alpha}) \geq 3$; obviously k depends only on α, and $r_k \geq 3r$. Then $|x-x_0|+2r < 3r \leq r_k$ by (11.102), hence $B(x_0, 2r) \subset B_k$ and $\omega^X(\Gamma \setminus B_k) \leq \omega^X(\Gamma \setminus B(x_0, 2r))$ because the harmonic measure is monotone. Together with (11.103) and (11.101), this proves that $\omega^X(\Gamma \setminus B(x_0, r)) \leq C^k \omega^X(\Gamma \setminus B(x_0, 2r))$, and (11.100) follows because k depends only on α. The proof of (11.99) would be similar.

LEMMA 11.14 (Comparison principle for global solutions). *Let $x_0 \in \Gamma$ and $r > 0$ be given, and let $X_0 := A_r(x_0) \in \Omega$ be the point given in Lemma 11.6. Let $u, v \in W$ be two non-negative, non identically zero, solutions of $Lu = Lv = 0$ in Ω such that $Tu = Tv = 0$ on $\Gamma \setminus B(x_0, r)$. Then*

$$C^{-1}\frac{u(X_0)}{v(X_0)} \leq \frac{u(X)}{v(X)} \leq C\frac{u(X_0)}{v(X_0)} \quad \text{for } X \in \Omega \setminus B(x_0, 2r), \tag{11.104}$$

where $C > 0$ depends only on n, d, C_0 and C_1.

REMARK 11.15. We also have (11.104) for any $X \in \Omega \setminus B(x_0, \alpha r)$, where $\alpha > 1$. In this case, the constant C depends also on α. We let the reader check that the proof below can be easily adapted to prove this too.

PROOF. By symmetry and as before, it is enough to prove that

$$\frac{u(X)}{v(X)} \leq C\frac{u(X_0)}{v(X_0)} \quad \text{for } X \in \Omega \setminus B(x_0, 2r). \tag{11.105}$$

Notice also that thanks to the Harnack inequality (Lemma 8.9), $v(X) > 0$ on the whole $\Omega \setminus \overline{B(x_0, r)}$, so we we don't need to be careful when we divide by $v(X)$.

Set $\Gamma_1 := \Gamma \cap B(x_0, r)$ and $\Gamma_2 := \Gamma \cap B(x_0, \frac{15}{8}r)$. Lemma 11.12 - or more exactly (11.99) - gives the following fact that will be of use later on:

$$\omega^X(\Gamma_2) \leq C\omega^X(\Gamma_1) \qquad \forall X \in \Omega \setminus B(x_0, 2r). \tag{11.106}$$

with a constant $C > 0$ which depends only on d, n, C_0 and C_1.

We claim that

$$v(X) \geq C^{-1}\omega^X(\Gamma_1)v(X_0) \quad \text{for } X \in \Omega \setminus B(x_0, 2r). \tag{11.107}$$

Indeed, by Harnack's inequality (Lemma 8.9),

$$v(X) \geq C^{-1}v(X_0) \quad \text{for } X \in B(X_0, \delta(X_0)/2). \tag{11.108}$$

Together with Lemma 11.4, which states that $g(X, X_0) \leq C\delta(X_0)^{1-d} \leq Cr^{1-d}$ for any $X \in \Omega \setminus B(X_0, \delta(X_0)/4)$, we deduce the existence of $K_1 > 0$ (that depends only on d, n, C_0 and C_1) such that

$$v(X) \geq K_1^{-1}r^{d-1}v(X_0)g(X, X_0) \quad \text{for } X \in B(X_0, \frac{1}{2}\delta(X_0)) \setminus B(X_0, \frac{1}{4}\delta(X_0)) \tag{11.109}$$

Let us apply the maximum principle (Lemma 11.3, with $E = \mathbb{R}^n \setminus \overline{B(X_0, \delta(X_0)/4)}$ and $F = \mathbb{R}^n \setminus B(X_0, \delta(X_0/2)))$, to the function $X \to v(X) - K_1^{-1} r^{d-1} v(X_0) g(X, X_0)$. The assumptions are satisfied because of (11.109), the properties of the Green function given in Lemma 10.1, and the fact that $v \in W$ is a non-negative solution of $Lv = 0$ on Ω. We get that

(11.110)
$$v(X) \geq K_1^{-1} r^{d-1} v(X_0) g(X, X_0) \quad \text{for } X \in \Omega \setminus B(X_0, \frac{1}{4}\delta(X_0)) \supset \Omega \setminus B(x_0, 2r).$$

The claim (11.107) is now a straightforward consequence of (11.110) and Lemma 11.11.

We want to prove now that

$$u(X) \leq C u(X_0) \omega^X(\Gamma_2) \quad \text{for } X \in \Omega \setminus B(x_0, 2r). \tag{11.111}$$

First, we need to prove that

$$u(X) \leq C u(X_0) \quad \text{for } X \in \left[B(x_0, \frac{13}{8}r) \setminus B(x_0, \frac{11}{8}r)\right] \cap \Omega. \tag{11.112}$$

We split $\left[B(x_0, \frac{13}{8}r) \setminus B(x_0, \frac{11}{8}r)\right] \cap \Omega$ into two sets:

$$\Omega_1 := \Omega \cap \{X \in B(x_0, \frac{13}{8}r) \setminus B(x_0, \frac{11}{8}r),\ \delta(X) < \frac{1}{8}r\} \tag{11.113}$$

and

$$\Omega_2 := \{X \in B(x_0, \frac{13}{8}r) \setminus B(x_0, \frac{11}{8}r),\ \delta(X) \geq \frac{1}{8}r\}. \tag{11.114}$$

The proof of (11.112) for $X \in \Omega_2$ is a consequence of the existence of Harnack chain (Lemma 2.2) and the Harnack inequality (Lemma 8.9). So it remains to prove (11.112) for $X \in \Omega_1$. Let thus $X \in \Omega_1$ be given. We can find $x \in \Gamma$ such that $X \in B(x, \frac{1}{8}r)$. Notice that $x \in B(x_0, \frac{7}{4}r)$ because $X \in B(x_0, \frac{13}{8}r)$. Yet, since u is a non-negative solution of $Lu = 0$ in $B(x, \frac{1}{4}r) \cap \Omega$ satisfying $Tu = 0$ on $B(x, \frac{1}{4}r) \cap \Gamma$, Lemma 11.8 gives that $u(Y) \leq C u(A_{r/8}(x))$ for $Y \in B(x, \frac{1}{8}r)$ and thus in particular $u(X) \leq C u(A_{r/8}(x))$. By the existence of Harnack chains (Lemma 2.2) and the Harnack inequality (Lemma 8.9) again, $u(A_{r/8}(x)) \leq C u(X_0)$. The bound (11.112) for all $X \in \Omega_1$ follows.

We proved (11.112) and now we want to get (11.111). Recall from Lemma 11.10 that $\omega^X(B(x_0, \frac{7}{4}r) \cap \Gamma) \geq C^{-1}$ for $X \in B(x_0, \frac{13}{8}r) \setminus \Gamma$. Hence, by (11.112),

(11.115)
$$u(X) \leq C u(X_0) \omega^X(B(x_0, \frac{7}{4}r) \cap \Gamma) \quad \text{for } X \in \left[B(x_0, \frac{13}{8}r) \setminus B(x_0, \frac{11}{8}r)\right] \cap \Omega.$$

Let $h \in C_0^\infty(B(x_0, \frac{15}{8}r))$ be such that $0 \leq h \leq 1$ and $h \equiv 1$ on $B(x_0, \frac{7}{4}r)$. Then let $u_h \in W$ be the solution of $Lu_h = 0$ with the Dirichlet condition $Tu_h = Th$. By the positivity of the harmonic measure,

$$u(X) \leq C u(X_0) u_h(X) \quad \text{for } X \in \left[B(x_0, \frac{13}{8}r) \setminus B(x_0, \frac{11}{8}r)\right] \cap \Omega. \tag{11.116}$$

The maximum principle given by Lemma 11.3 - where we take $E = \mathbb{R}^n \setminus \overline{B(x_0, \frac{11}{8}r)}$ and $F = \mathbb{R}^n \setminus B(x_0, \frac{13}{8}r)$ - yields

$$u(X) \leq C u(X_0) u_h(X) \quad \text{for } X \in \Omega \setminus B(x_0, \frac{13}{8}r) \tag{11.117}$$

and hence

$$u(X) \leq Cu(X_0)\omega^X(\Gamma_2) \quad \text{for } X \in \Omega \setminus B(x_0, \frac{13}{8}r), \tag{11.118}$$

where we use again the positivity of the harmonic measure. The assertion (11.111) is now proven.

We conclude the proof of the lemma by gathering the previous results. Because of (11.107) and (11.111),

$$\frac{u(X)}{v(X)} \leq C\frac{u(X_0)}{v(X_0)}\frac{\omega^X(\Gamma_2)}{\omega^X(\Gamma_1)} \quad \text{for } X \in \Omega \setminus B(x_0, 2r), \tag{11.119}$$

and (11.105) follows from (11.106). Lemma 11.14 follows. □

Note that the functions $X \to \omega^X(E)$, where $E \subset \Gamma$ is a non-trivial Borel set, do not lie in W and thus cannot be used directly in Lemma 11.14. The following lemma solves this problem.

LEMMA 11.16 (Comparison principle for harmonic measures / Change of poles). *Let $x_0 \in \Gamma$ and $r > 0$ be given, and let $X_0 := A_r(x_0) \in \Omega$ be as in Lemma 11.6. Let $E, F \subset \Gamma \cap B(x_0, r)$ be two Borel subsets of Γ such that $\omega^{X_0}(E)$ and $\omega^{X_0}(F)$ are positive. Then*

$$C^{-1}\frac{\omega^{X_0}(E)}{\omega^{X_0}(F)} \leq \frac{\omega^X(E)}{\omega^X(F)} \leq C\frac{\omega^{X_0}(E)}{\omega^{X_0}(F)} \quad \text{for } X \in \Omega \setminus B(x_0, 2r), \tag{11.120}$$

where $C > 0$ depends only on n, d, C_0 and C_1. In particular, with the choice $F = B(x_0, r) \cap \Gamma$,

$$C^{-1}\omega^{X_0}(E) \leq \frac{\omega^X(E)}{\omega^X(B(x_0, r) \cap \Gamma)} \leq C\omega^{X_0}(E) \quad \text{for } X \in \Omega \setminus B(x_0, 2r), \tag{11.121}$$

where again $C > 0$ depends only on n, d, C_0 and C_1.

PROOF. The second part of the lemma, that is (11.121) an immediate consequence of (11.120) and the non-degeneracy of the harmonic measure (Lemma 11.10). In addition, it is enough to prove

$$C^{-1}\frac{\omega^{X_0}(E)}{u(X_0)} \leq \frac{\omega^X(E)}{u(X)} \leq C\frac{\omega^{X_0}(E)}{u(X_0)}, \tag{11.122}$$

where $u \in W$ is any non-negative non-zero solution of $Lu = 0$ in Ω satisfying $Tu = 0$ on $\Gamma \setminus B(x_0, r)$, and $C > 0$ depends only on n, d, C_0 and C_1. Indeed, (11.120) follows by applying (11.122) to both E and F. Incidentally, it is very easy to find u like this: just apply Lemma 9.4 to a smooth bump function g with a small compact support near x_0.

Assume first that $E = K$ is a compact set. Let $X \in \Omega \setminus B(x_0, 2r)$ be given. Thanks to Lemma 9.7 (i), the assumption $\omega^{X_0}(K) > 0$ implies that $\omega^X(K) > 0$. By the the regularity of the harmonic measure (see (9.27)), we can find an open set $U_X \supset K$ such that

$$\omega^{X_0}(U_X) \leq 2\omega^{X_0}(K) \text{ and } \omega^X(U_X) \leq 2\omega^X(K). \tag{11.123}$$

Urysohn's lemma (see Lemma 2.12 in [**Rud**]) gives a function $h \in C_0^0(\Gamma)$ such that $\mathbb{1}_K \leq h \leq \mathbb{1}_{U_X}$. Write $v^h = U(h)$ for the image of the function h by the map given in Lemma 9.4. We have seen for the proof of Lemma 9.4 that h can be approximated, in the supremum norm, by smooth, compactly supported functions

h_k, and that the corresponding solutions $v_k = U(h_k)$, and that can also obtained through 9.1, lie in W and converge to v^h uniformly on Ω. Hence we can find $k > 0$ such that

$$\frac{1}{2} v_k \leq v^h \leq 2v_k \tag{11.124}$$

everywhere in Ω. Write v for v_k. Notice that v depends on X, but it has no importance. The estimates (11.123) and (11.124) give

$$\frac{1}{4} v(X_0) \leq \omega^{X_0}(K) \leq 2v(X_0) \quad \text{and} \quad \frac{1}{4} v(X) \leq \omega^X(K) \leq 2v(X). \tag{11.125}$$

We can even choose $U_X \supset K$ so small, and then g_k with a barely larger support, so that $Tv = g_k$ is supported in $B(x_0, r)$. As a consequence, the solution v satisfies the assumption of Lemma 11.14. Hence, the latter entails

$$C^{-1} \frac{v(X_0)}{u(X_0)} \leq \frac{v(X)}{u(X)} \leq C \frac{v(X_0)}{u(X_0)} \tag{11.126}$$

with a constant $C > 0$ that depends only on d, n, C_0 and C_1. Together with (11.125), we get that

$$C^{-1} \frac{\omega^{X_0}(K)}{u(X_0)} \leq \frac{\omega^X(K)}{u(X)} \leq C \frac{\omega^{X_0}(K)}{u(X_0)} \tag{11.127}$$

with a constant $C > 0$ that still depends only on d, n, C_0 and C_1 (and thus is independent of X). Thus the conclusion (11.120) holds whenever $E = K$ is a compact set.

Now let E be any Borel subset of $\Gamma \cap B(x_0, r)$. Let $X \in \Omega \setminus B(x_0, 2r)$. According to the regularity of the harmonic measure (9.27), there exists $K_X \subset E$ (depending on X) such that

$$\omega^{X_0}(K_X) \leq \omega^{X_0}(E) \leq 2\omega^{X_0}(K_X) \quad \text{and} \quad \omega^X(K_X) \leq \omega^X(E) \leq 2\omega^X(K_X). \tag{11.128}$$

The combination of (11.128) and (11.127) (applied to K_X) yields

$$C^{-1} \frac{\omega^{X_0}(E)}{u(X_0)} \leq \frac{\omega^X(E)}{u(X)} \leq C \frac{\omega^{X_0}(E)}{u(X_0)} \tag{11.129}$$

where the constant $C > 0$ depends only upon d, n, C_0 and C_1. The lemma follows. □

Let us prove now a comparison principle for the solution that are not defined in the whole domain Ω.

THEOREM 11.17 (Comparison principle for locally defined functions). *Let $x_0 \in \Gamma$ and $r > 0$ and let $X_0 := A_r(x_0) \in \Omega$ be the point given in Lemma 11.6. Let $u, v \in W_r(B(x_0, 2r))$ be two non-negative, not identically zero, solutions of $Lu = Lv = 0$ in $B(x_0, 2r)$, such that $Tu = Tv = 0$ on $\Gamma \cap B(x_0, 2r)$. Then*

$$C^{-1} \frac{u(X_0)}{v(X_0)} \leq \frac{u(X)}{v(X)} \leq C \frac{u(X_0)}{v(X_0)} \quad \text{for } X \in \Omega \cap B(x_0, r), \tag{11.130}$$

where $C > 0$ depends only on n, d, C_0 and C_1.

PROOF. The plan of the proof is as follows: first, for $y_0 \in \Gamma$ and $s > 0$, we construct a function $f_{y_0,s}$ on Ω such that (i) $f_{y_0,s}(X)$ is equivalent to $\omega^X(\Gamma \setminus B(y_0, 2s))$ when $X \in B(y_0, s)$ is close to Γ and (ii) $f_{y_0,s}(X)$ is negative when

$X \in \Omega \setminus B(y_0, Ms)$ - with M depending only on d, n, C_0 and C_1. We use $f_{y_0,s}$ to prove that $v(X) \geq v(A_s(y_0))\omega^X(\Gamma \setminus B(y_0, 2s))$ whenever $X \in B(y_0, s)$ and $B(y_0, Ms) \subset B(x_0, 2r)$ is a ball centered on Γ. We use then an appropriate covering of $B(x_0, r)$ by balls and the Harnack inequality to get the lower bound $v(X) \geq v(X_0)\omega^X(\Gamma \setminus B(x_0, 4r))$, which is the counterpart of (11.107) in our context. We conclude as in Lemma 11.14 by using Lemma 11.8 and the doubling property for the harmonic measure (Lemma 11.12)

Let $y_0 \in \Gamma$ and $s > 0$. Write Y_0 for $A_s(y_0)$. The main idea is to take

$$f_{y_0,s}(X) := s^{d-1}g(X, Y_0) - K_1\omega^X(\Gamma \setminus B(y_0, K_2 s)) \tag{11.131}$$

for some $K_1, K_2 > 0$ that depend only on n, d, C_0 and C_1. With good choices of K_1 and K_2, the function $f_{y_0,s}$ is positive in $B(y_0, s)$ and negative outside of a big ball $B(y_0, 2K_2 s)$. However, with this definition involving the harmonic measure, the function $f_{y_0,s}$ doesn't satisfy the appropriate estimates required for the use of the maximum principle given as Lemma 11.3. So we shall replace $\omega^X(\Gamma \setminus B(y_0, K_2 s))$ by some solution of $Lu = 0$, with smooth Dirichlet condition.

Let $h \in C^\infty(\mathbb{R}^n)$ be such that $0 \leq h \leq 1$, $h \equiv 0$ on $B(0, 1/2)$ and $h \equiv 1$ on the complement of $B(0,1)$. For $\beta > 0$ (which will be chosen large), we define h_β by $h_\beta(x) = h(\frac{x-y_0}{\beta s})$. Let u_β be the solution, given by Lemma 9.1, of $Lu_\beta = 0$ with the Dirichlet condition $Tu_\beta = Th_\beta$. Notice that $u_\beta \in W$ because $1 - u_\beta$ is the solution of L with the smooth and compactly supported trace $1 - h$. Observe that for any $X \in \Omega$ and $\beta > 0$,

$$\omega^X(\Gamma \setminus B(y_0, \beta s)) \leq u_\beta(X) \leq \omega^X(\Gamma \setminus B(y_0, \beta s/2)). \tag{11.132}$$

The functions u_β will play the role of harmonic measures but, unlike these, the functions u_β lie in W and are thus suited for the use of Lemma 11.3.

By Lemma 11.4, there exists $C > 0$, that depends only on d, n, C_0 and C_1, such that

$$g(X, Y_0) \leq C\delta(Y_0)^{1-d} \quad \text{for } X \in \Omega \setminus B(Y_0, \delta(Y_0)/4). \tag{11.133}$$

Moreover, since Y_0 comes from Lemma 11.6, we have $\epsilon s \leq \delta(Y_0) \leq s$ with an $\epsilon > 0$ that does not depend on s or y_0, and hence

$$s^{d-1}g(X, Y_0) \leq C \quad \text{for } X \in \Omega \setminus B(y_0, 2s). \tag{11.134}$$

From this and the non-degeneracy of the harmonic measure (Lemma 11.10), we deduce that for $\beta \geq 1$,

$$s^{d-1}g(X, Y_0) \leq K_1\omega^X(\Gamma \setminus B(y_0, \beta s)) \leq K_1 u_\beta(X) \quad \text{for } X \in \Omega \setminus B(y_0, 2\beta s), \tag{11.135}$$

where the constant $K_1 > 0$, depends only on d, n, C_0 and C_1.

Our aim now is to find $K_2 \geq 20$ such that

$$K_1 u_{K_2}(X) \leq \frac{1}{2}s^{d-1}g(X, Y_0) \quad \text{for } X \in \Omega \cap [B(y_0, s) \setminus B(Y_0, \delta(Y_0)/4)]. \tag{11.136}$$

By construction, the function u_β is such that $T(u_\beta) = 0$ on $\Gamma \cap B(y_0, \beta s/2)$, so according to the Hölder continuity at the boundary (Lemma 8.16), we have

$$\sup_{B(y_0,10s)} u_\beta \leq C\beta^{-\alpha} \tag{11.137}$$

for any $\beta \geq 20$, where C and $\alpha > 0$ depend only on d, n, C_0 and C_1. Moreover, due to (11.132) and the non-degeneracy of the harmonic measure (Lemma 11.10),

$$u_4(X) \geq C^{-1} \quad \text{for } X \in \Omega \setminus B(y_0, 8s) \tag{11.138}$$

where u_4 is defined as u_β (with $\beta = 4$). As a consequence, there exists $K_3 > 0$, that depends only on d, n, C_0, and C_1, such that for $\beta \geq 20$,

$$u_\beta(X) \leq K_3 \beta^{-\alpha} u_4(X) \quad \text{for } X \in \Omega \cap [B(y_0, 10s) \setminus B(y_0, 8s)]. \tag{11.139}$$

We just proved that for $\beta \geq 20$, the function $u' = K_3\beta^{-\alpha}u_4 - u_\beta$ satisfies the assumption (iii) of Lemma 11.3, with $E = B(y_0, 10s)$ and $F = \overline{B(y_0, 8s)}$. The other assumptions of Lemma 11.3 are satisfied as well, since $u' \in W$ is smooth and $T(u') = K_3\beta^{-\alpha}Tu_4 \geq 0$ on $\Gamma \cap E$. Therefore, Lemma 11.3 gives

$$u_\beta(X) \leq K_3 \beta^{-\alpha} u_4(X) \quad \text{for } X \in \Omega \cap B(y_0, 10s). \tag{11.140}$$

Use now (11.132) and Lemma 11.11 to get for $X \in \Omega \cap [B(y_0, s) \setminus B(Y_0, \delta(Y_0)/4)]$,

$$u_\beta(X) \leq K_3 \beta^{-\alpha} \omega^X(\Gamma \setminus B(y_0, 2s)) \leq C\beta^{-\alpha} s^{d-1} g(X, Y_0), \tag{11.141}$$

where $C > 0$ depends only on d, n, C_0 and C_1. The existence of $K_2 \geq 20$ satisfying (11.136) is now immediate.

Define the function $f_{y_0,s}$ on $\Omega \setminus \{Y_0\}$ by

$$f_{y_0,s}(X) := s^{d-1} g(X, Y_0) - K_1 u_{K_2}(X). \tag{11.142}$$

The inequality (11.135) gives

$$f_{y_0,s}(X) \leq 0 \quad \text{for } X \in \Omega \setminus B(y_0, 2K_2 s), \tag{11.143}$$

and the estimates (11.136) and (11.69) imply that

$$f_{y_0,s}(X) \geq \frac{1}{2} s^{d-1} g(X, Y_0) \geq C^{-1} \omega^X(\Gamma \setminus B(y_0, 2s)) \tag{11.144}$$

for

$$X \in \Omega \cap [B(y_0, s) \setminus B(Y_0, \delta(Y_0)/4)].$$

Let us turn to the proof of the comparison principle. By symmetry and as in Lemma 11.14, it suffices to prove the upper bound in (11.130), that is

$$\frac{u(X)}{v(X)} \leq C \frac{u(X_0)}{v(X_0)} \quad \text{for } X \in \Omega \cap B(x_0, r). \tag{11.145}$$

We claim that

$$v(X) \geq C^{-1} v(X_0) \omega^X(\Gamma \setminus B(x_0, 2r)) \quad \text{for } X \in \Omega \cap B(x_0, r), \tag{11.146}$$

where $C > 0$ depends only on n, d, C_0 and C_1. So let $X \in \Omega \cap B(x_0, r)$ be given. Two cases may happen. If $\delta(X) \geq \frac{r}{8K_2}$, where K_2 comes from (11.136) and is the same as in the definition of $f_{y_0,s}$, the existence of Harnack chains (Lemma 2.2), the Harnack inequality (Lemma 8.9) and the non-degeneracy of the harmonic measure (Lemma 11.10) give

$$v(X) \approx v(X_0) \approx v(X_0) \frac{\omega^X(\Gamma \setminus B(x_0, 2r))}{\omega^{X_0}(\Gamma \setminus B(x_0, 2r))} \approx v(X_0) \omega^X(\Gamma \setminus B(x_0, 2r)) \tag{11.147}$$

by (11.67). The more interesting remaining case is when $\delta(X) < \frac{r}{8K_2}$. Take $y_0 \in \Gamma$ such that $|X - y_0| = \delta(X)$. Set $s := \frac{r}{8K_2}$ and $Y_0 = A_s(y_0)$. The ball $B(y_0, \frac{1}{2}r) = B(y_0, 4K_2 s)$ is contained in $B(x_0, \frac{7}{4}r)$. The following points hold :

- The quantity $\int_{B(y_0,4K_2s)\setminus B(Y_0,\delta(Y_0)/4)} |\nabla v|^2 dm$ is finite because
$$v \in W_r(B(x_0, 2r)).$$
The fact that $\int_{B(y_0,4K_2s)\setminus B(Y_0,\delta(Y_0)/4)} |\nabla f_{y_0,s}|^2 dm$ is finite as well follows from the property (10.2) of the Green function.
- There exists $K_4 > 0$ (depending only on d, n, C_0 and C_1) such that
$$v(Y) - K_4 v(Y_0) f_{y_0,s}(Y) \geq 0 \quad \text{for } Y \in B(Y_0, \delta(Y_0)/2) \setminus B(Y_0, \delta(Y_0)/4). \tag{11.148}$$
This latter inequality is due to the following two bounds: the fact that
$$f_{y_0,s}(Y) \leq s^{1-d} g(Y, Y_0) \leq C \quad \text{for } Y \in B(Y_0, \delta(Y_0)/2) \setminus B(Y_0, \delta(Y_0)/4), \tag{11.149}$$
which is a consequence of the definition (11.142) and (10.6), and the bound
$$v(Y) \geq C^{-1} v(Y_0) \quad \text{for } Y \in B(Y_0, \delta(Y_0)/2), \tag{11.150}$$
which comes from the Harnack inequality (Lemma 8.9).
- The function $v - K_4 v(Y_0) f_{y_0,s}$ is nonnegative on
$$\Omega \cap [B(y_0, 4K_2s) \setminus B(y_0, 2K_2s)].$$
Indeed, $v \geq 0$ on $B(y_0, 4K_2s)$ and, thanks to (11.143), $f_{y_0,s} \leq 0$ on $\Omega \setminus B(y_0, 2K_2s)$.
- The trace of $v - K_4 v(Y_0) f_{y_0,s}$ is non-negative on $B(y_0, 4K_2s) \cap \Gamma$ again because $Tv = 0$ on $B(y_0, 4K_2s) \cap \Gamma$ and $T[f_{y_0,s}] \leq 0$ on $B(y_0, 4K_2s) \cap \Gamma$ by construction.

The previous points prove that $v - K_4 v(Y_0) f_{y_0,s}$ satisfies the assumptions of Lemma 11.3 with $E = B(y_0, 4K_2s) \setminus \overline{B(Y_0, \delta(Y_0)/4)}$ and $F = \overline{B(y_0, 2K_2s)} \setminus B(Y_0, \delta(Y_0)/2)$. As a consequence, for any $Y \in B(y_0, 4K_2s) \setminus B(Y_0, \delta(Y_0)/4)$
$$v(Y) - K_4 v(Y_0) f_{y_0,s}(Y) \geq 0, \tag{11.151}$$
and hence, for any $Y \in B(y_0, s) \setminus B(Y_0, \delta(Y_0)/4)$
$$v(Y) \geq K_4 v(Y_0) f_{y_0,s}(Y) \geq C^{-1} v(Y_0) \omega^Y(\Gamma \setminus B(y_0, 2s)) \tag{11.152}$$
by (11.144). Since both v and $Y \to \omega^Y(\Gamma \setminus B(y_0, 2s))$ are solutions on $B(y_0, 2s)$, we can use the Harnack inequality (Lemma 8.9) to deduce, first, that (11.152) holds for any $Y \in B(y_0, s)$ and second, that we can replace $v(Y_0)$ by $v(X_0)$ (recall that at this point, $\frac{s}{r} = \frac{1}{8K_2}$ is controlled by the usual constants). Therefore,
$$v(Y) \geq C^{-1} v(X_0) \omega^Y(\Gamma \setminus B(y_0, 2s)) \quad \text{for } Y \in B(y_0, s). \tag{11.153}$$
In particular, with our choice of y_0 and s, the inequality is true when $X = Y$, that is,
$$v(X) \geq C^{-1} v(X_0) \omega^X(\Gamma \setminus B(y_0, 2s)) \geq C^{-1} v(X_0) \omega^X(\Gamma \setminus B(x_0, 2r)) \tag{11.154}$$
where $C > 0$ depends only on d, n, C_0 and C_1. The claim (11.146) follows.

Now we want to prove that
$$u(X) \leq C u(X_0) \omega^X(\Gamma \setminus B(x_0, \frac{5}{4}r)) \quad \text{for } X \in \Omega \cap B(x_0, r). \tag{11.155}$$
By Lemma 11.8,
$$u(X) \leq C u(X_0) \quad \text{for } X \in \Omega \cap B(x_0, \frac{7}{4}r). \tag{11.156}$$

Pick $h' \in C^\infty(\mathbb{R}^n)$ such that $0 \leq h' \leq 1$, $h' \equiv 1$ outside of $B(x_0, \frac{3}{2}r)$, and $h' \equiv 0$ on $B(x_0, \frac{5}{4}r)$. Let $u_{h'} = U(h')$ be the solution of $Lu_{h'} = 0$ with the data $Tu_{h'} = Th'$ (given by Lemma 9.1). As before, $u_{h'} \in W$ because $1 - u_{h'} = U(1-h)$ and $1-h$ is a test function. Also, $u_{h'}(X) \geq \omega^X(\Gamma \setminus B(x_0, \frac{3}{2}r))$ by monotonicity. So (11.66), which states the non-degeneracy of the harmonic measure, gives

$$u_{h'}(X) \geq C^{-1} \quad \text{for } X \in \Omega \setminus B(x_0, \frac{13}{8}r). \tag{11.157}$$

The combination of (11.156) and (11.157) yields the existence of $K_5 > 0$ (that depends only on d, n, C_0 and C_1) such that $K_5 u(X_0) u_{h'} - u \geq 0$ on $\Omega \cap [B(x_0, \frac{7}{4}r) \setminus B(x_0, \frac{13}{8}r)]$. It is easy to check that $K_5 u(X_0) u_{h'} - u$ satisfies all the assumptions of Lemma 11.3, with $E = B(x_0, \frac{7}{4}r)$ and $F = \overline{B(x_0, \frac{13}{8}r)}$. This is because $u \in W_r(B(x_0, 2r))$, $u_{h'} \in W$, $Tu_{h'} \geq 0$, and $Tu = 0$ on $\Gamma \cap B(x_0, 2r)$. Then by Lemma 11.3

$$u \leq K_5 u(X_0) u_{h'} \quad \text{for } X \in \Omega \cap B(x_0, \frac{7}{4}r), \tag{11.158}$$

and since $u_{h'}(X) \leq \omega^X(\Gamma \setminus B(x_0, \frac{5}{4}r))$ for all $X \in \Omega$,

$$u(X) \leq C u(X_0) \omega^X(\Gamma \setminus B(x_0, \frac{5}{4}r)) \quad \text{for } X \in \Omega \cap B(x_0, \frac{7}{4}r). \tag{11.159}$$

The claim (11.155) follows.

The bounds (11.146) and (11.155) imply that

$$\frac{u(X)}{v(X)} \leq C \frac{u(X_0)}{v(X_0)} \frac{\omega^X(\Gamma \setminus B(x_0, \frac{5}{4}r))}{\omega^X(\Gamma \setminus B(x_0, 2r))} \quad \text{for } X \in \Omega \cap B(x_0, r). \tag{11.160}$$

The bound (11.145) is now a consequence of the above inequality and the doubling property of the harmonic measure (Lemma 11.12, or more exactly (11.100)). □

Bibliography

[A] A. Ancona, *Théorie du potentiel sur les graphes et les variétés* (French), École d'été de Probabilités de Saint-Flour XVIII—1988, Lecture Notes in Math., vol. 1427, Springer, Berlin, 1990, pp. 1–112, DOI 10.1007/BFb0103041. MR1100282

[AGMT] J. Azzam, J. Garnett, M. Mourgoglou, X. Tolsa. *Uniform rectifiability, elliptic measure, square functions, and ε-approximability via an ACF monotonicity formula*, preprint.

[AHMNT] J. Azzam, S. Hofmann, J. M. Martell, K. Nyström, and T. Toro, *A new characterization of chord-arc domains*, J. Eur. Math. Soc. (JEMS) **19** (2017), no. 4, 967–981, DOI 10.4171/JEMS/685. MR3626548

[AHM3TV] J. Azzam, S. Hofmann, J. M. Martell, S. Mayboroda, M. Mourgoglou, X. Tolsa, and A. Volberg, *Rectifiability of harmonic measure*, Geom. Funct. Anal. **26** (2016), no. 3, 703–728, DOI 10.1007/s00039-016-0371-x. MR3540451

[Ba] M. Badger, *Null sets of harmonic measure on NTA domains: Lipschitz approximation revisited*, Math. Z. **270** (2012), no. 1-2, 241–262, DOI 10.1007/s00209-010-0795-1. MR2875832

[BJ] C. J. Bishop and P. W. Jones, *Harmonic measure and arclength*, Ann. of Math. (2) **132** (1990), no. 3, 511–547, DOI 10.2307/1971428. MR1078268

[CFK] L. A. Caffarelli, E. B. Fabes, and C. E. Kenig, *Completely singular elliptic-harmonic measures*, Indiana Univ. Math. J. **30** (1981), no. 6, 917–924, DOI 10.1512/iumj.1981.30.30067. MR632860

[CFMS] L. Caffarelli, E. Fabes, S. Mortola, and S. Salsa, *Boundary behavior of nonnegative solutions of elliptic operators in divergence form*, Indiana Univ. Math. J. **30** (1981), no. 4, 621–640, DOI 10.1512/iumj.1981.30.30049. MR620271

[Da] B. E. J. Dahlberg, *On the absolute continuity of elliptic measures*, Amer. J. Math. **108** (1986), no. 5, 1119–1138, DOI 10.2307/2374598. MR859772

[DJ] G. David and D. Jerison, *Lipschitz approximation to hypersurfaces, harmonic measure, and singular integrals*, Indiana Univ. Math. J. **39** (1990), no. 3, 831–845, DOI 10.1512/iumj.1990.39.39040. MR1078740

[DFM] G. David, J. Feneuil, and S. Mayboroda, *Harmonic measure on sets of codimension larger than one* (English, with English and French summaries), C. R. Math. Acad. Sci. Paris **355** (2017), no. 4, 406–410, DOI 10.1016/j.crma.2017.02.013. MR3634676

[DK] H. Dong and S. Kim, *Green's matrices of second order elliptic systems with measurable coefficients in two dimensional domains*, Trans. Amer. Math. Soc. **361** (2009), no. 6, 3303–3323, DOI 10.1090/S0002-9947-09-04805-3. MR2485428

[Duo] J. Duoandikoetxea, *Fourier analysis*, Graduate Studies in Mathematics, vol. 29, American Mathematical Society, Providence, RI, 2001. Translated and revised from the 1995 Spanish original by David Cruz-Uribe. MR1800316

[Eva] L. C. Evans, *Partial differential equations*, 2nd ed., Graduate Studies in Mathematics, vol. 19, American Mathematical Society, Providence, RI, 2010, DOI 10.1090/gsm/019. MR2597943

[FKS] E. B. Fabes, C. E. Kenig, and R. P. Serapioni, *The local regularity of solutions of degenerate elliptic equations*, Comm. Partial Differential Equations **7** (1982), no. 1, 77–116, DOI 10.1080/03605308208820218. MR643158

[FJK] E. Fabes, D. Jerison, and C. Kenig, *The Wiener test for degenerate elliptic equations* (English, with French summary), Ann. Inst. Fourier (Grenoble) **32** (1982), no. 3, vi, 151–182. MR688024

[Fed] H. Federer, *Geometric measure theory*, Die Grundlehren der mathematischen Wissenschaften, Band 153, Springer-Verlag New York Inc., New York, 1969. MR0257325

[GMT] J. Garnett, M. Mourgoglou, and X. Tolsa, *Uniform rectifiability from Carleson measure estimates and ε-approximability of bounded harmonic functions*, Duke Math. J. **167** (2018), no. 8, 1473–1524, DOI 10.1215/00127094-2017-0057. MR3807315

[GT] D. Gilbarg and N. S. Trudinger, *Elliptic partial differential equations of second order*, Classics in Mathematics, Springer-Verlag, Berlin, 2001. Reprint of the 1998 edition. MR1814364

[GW] M. Grüter and K.-O. Widman, *The Green function for uniformly elliptic equations*, Manuscripta Math. **37** (1982), no. 3, 303–342, DOI 10.1007/BF01166225. MR657523

[Haj] P. Hajlasz, *Geometric approach to Sobolev spaces and badly degenerated elliptic equations*, Nonlinear analysis and applications (Warsaw, 1994), GAKUTO Internat. Ser. Math. Sci. Appl., vol. 7, Gakkōtosho, Tokyo, 1996, pp. 141–168. MR1422932

[HaK] P. Hajłasz and P. Koskela, *Sobolev meets Poincaré* (English, with English and French summaries), C. R. Acad. Sci. Paris Sér. I Math. **320** (1995), no. 10, 1211–1215. MR1336257

[HaK2] ——— *Sobolev met Poincaré*. Mem. Amer. Math. Soc., 145 (2000), no. 688.

[HL] Q. Han and F. Lin, *Elliptic partial differential equations*, 2nd ed., Courant Lecture Notes in Mathematics, vol. 1, Courant Institute of Mathematical Sciences, New York; American Mathematical Society, Providence, RI, 2011. MR2777537

[HKM] J. Heinonen, T. Kilpeläinen, and O. Martio, *Nonlinear potential theory of degenerate elliptic equations*, Oxford Mathematical Monographs, The Clarendon Press, Oxford University Press, New York, 1993. Oxford Science Publications. MR1207810

[HoK] S. Hofmann and S. Kim, *The Green function estimates for strongly elliptic systems of second order*, Manuscripta Math. **124** (2007), no. 2, 139–172, DOI 10.1007/s00229-007-0107-1. MR2341783

[HM1] S. Hofmann and J. M. Martell, *Uniform rectifiability and harmonic measure I: Uniform rectifiability implies Poisson kernels in L^p* (English, with English and French summaries), Ann. Sci. Éc. Norm. Supér. (4) **47** (2014), no. 3, 577–654, DOI 10.24033/asens.2223. MR3239100

[HLMN] S. Hofmann, P. Le, J. M. Martell, and K. Nyström, *The weak-A_∞ property of harmonic and p-harmonic measures implies uniform rectifiability*, Anal. PDE **10** (2017), no. 3, 513–558, DOI 10.2140/apde.2017.10.513. MR3641879

[HMM1] S. Hofmann, J. M. Martell, and S. Mayboroda, *Uniform rectifiability, Carleson measure estimates, and approximation of harmonic functions*, Duke Math. J. **165** (2016), no. 12, 2331–2389, DOI 10.1215/00127094-3477128. MR3544283

[HMM2] ——— *Transference of scale-invariant estimates from Lipschitz to Non-tangentially accessible to Uniformly rectifiable domains,* preprint.

[HMU] S. Hofmann, J. M. Martell, and I. Uriarte-Tuero, *Uniform rectifiability and harmonic measure, II: Poisson kernels in L^p imply uniform rectifiability*, Duke Math. J. **163** (2014), no. 8, 1601–1654, DOI 10.1215/00127094-2713809. MR3210969

[JW] A. Jonsson and H. Wallin, *Function spaces on subsets of* $\mathbf{R}^n$, Math. Rep. **2** (1984), no. 1, xiv+221. MR820626

[Ken] C. E. Kenig, *Harmonic analysis techniques for second order elliptic boundary value problems*, CBMS Regional Conference Series in Mathematics, vol. 83, Published for the Conference Board of the Mathematical Sciences, Washington, DC; by the American Mathematical Society, Providence, RI, 1994, DOI 10.1090/cbms/083. MR1282720

[KKPT] C. Kenig, H. Koch, J. Pipher, and T. Toro, *A new approach to absolute continuity of elliptic measure, with applications to non-symmetric equations*, Adv. Math. **153** (2000), no. 2, 231–298, DOI 10.1006/aima.1999.1899. MR1770930

[KKiPT] C. Kenig, B. Kirchheim, J. Pipher, and T. Toro, *Square functions and the A_∞ property of elliptic measures*, J. Geom. Anal. **26** (2016), no. 3, 2383–2410, DOI 10.1007/s12220-015-9630-6. MR3511480

[KJ] D. S. Jerison and C. E. Kenig, *Boundary behavior of harmonic functions in non-tangentially accessible domains*, Adv. in Math. **46** (1982), no. 1, 80–147, DOI 10.1016/0001-8708(82)90055-X. MR676988

[KP] C. E. Kenig and J. Pipher, *The Dirichlet problem for elliptic equations with drift terms*, Publ. Mat. **45** (2001), no. 1, 199–217, DOI 10.5565/PUBLMAT_45101_09. MR1829584

[Kil] T. Kilpeläinen, *Weighted Sobolev spaces and capacity*, Ann. Acad. Sci. Fenn. Ser. A I Math. **19** (1994), no. 1, 95–113. MR1246890

[Lv] M. Lavrent′ev, *Boundary problems in the theory of univalent functions*, Amer. Math. Soc. Transl. (2) **32** (1963), 1–35. MR0155970

[LN] J. L. Lewis and K. Nyström, *Quasi-linear PDEs and low-dimensional sets*, J. Eur. Math. Soc. (JEMS) **20** (2018), no. 7, 1689–1746, DOI 10.4171/JEMS/797. MR3807311

[MZ] J. Malý and W. P. Ziemer, *Fine regularity of solutions of elliptic partial differential equations*, Mathematical Surveys and Monographs, vol. 51, American Mathematical Society, Providence, RI, 1997, DOI 10.1090/surv/051. MR1461542

[Maz] V. Maz'ya, *Sobolev spaces with applications to elliptic partial differential equations*, Second, revised and augmented edition, Grundlehren der Mathematischen Wissenschaften [Fundamental Principles of Mathematical Sciences], vol. 342, Springer, Heidelberg, 2011, DOI 10.1007/978-3-642-15564-2. MR2777530

[RR] F. & M. Riesz. *Über die randwerte einer analtischen funktion.* Compte Rendues du Quatrième Congrès des Mathématiciens Scandinaves, Stockholm 1916, Almqvists and Wilksels, Upsala, 1920.

[Rud] W. Rudin, *Real and complex analysis*, McGraw-Hill Book Co., New York-Toronto, Ont.-London, 1966. MR0210528

[Se] S. Semmes, *Analysis vs. geometry on a class of rectifiable hypersurfaces in* $\mathbf{R}^n$, Indiana Univ. Math. J. **39** (1990), no. 4, 1005–1035, DOI 10.1512/iumj.1990.39.39048. MR1087183

[Sta] G. Stampacchia, *Le problème de Dirichlet pour les équations elliptiques du second ordre à coefficients discontinus* (French), Ann. Inst. Fourier (Grenoble) **15** (1965), no. fasc. 1, 189–258. MR0192177

[Sta2] G. Stampacchia, *Formes bilinéaires coercitives sur les ensembles convexes* (French), C. R. Acad. Sci. Paris **258** (1964), 4413–4416. MR0166591

[Ste] E. M. Stein, *Singular integrals and differentiability properties of functions*, Princeton Mathematical Series, No. 30, Princeton University Press, Princeton, N.J., 1970. MR0290095

[Wu] J.-M. Wu, *On singularity of harmonic measure in space*, Pacific J. Math. **121** (1986), no. 2, 485–496. MR819202

[Z] W. P. Ziemer, *Some remarks on harmonic measure in space*, Pacific J. Math. **55** (1974), 629–637. MR0427657

Editorial Information

To be published in the *Memoirs*, a paper must be correct, new, nontrivial, and significant. Further, it must be well written and of interest to a substantial number of mathematicians. Piecemeal results, such as an inconclusive step toward an unproved major theorem or a minor variation on a known result, are in general not acceptable for publication.

Papers appearing in *Memoirs* are generally at least 80 and not more than 200 published pages in length. Papers less than 80 or more than 200 published pages require the approval of the Managing Editor of the Transactions/Memoirs Editorial Board. Published pages are the same size as those generated in the style files provided for $\mathcal{A}\mathcal{M}\mathcal{S}$-LaTeX or $\mathcal{A}\mathcal{M}\mathcal{S}$-TeX.

Information on the backlog for this journal can be found on the AMS website starting from `http://www.ams.org/memo`.

A Consent to Publish is required before we can begin processing your paper. After a paper is accepted for publication, the Providence office will send a Consent to Publish and Copyright Agreement to all authors of the paper. By submitting a paper to the *Memoirs*, authors certify that the results have not been submitted to nor are they under consideration for publication by another journal, conference proceedings, or similar publication.

Information for Authors

Memoirs is an author-prepared publication. Once formatted for print and on-line publication, articles will be published as is with the addition of AMS-prepared frontmatter and backmatter. Articles are not copyedited; however, confirmation copy will be sent to the authors.

Initial submission. The AMS uses Centralized Manuscript Processing for initial submissions. Authors should submit a PDF file using the Initial Manuscript Submission form found at `www.ams.org/submission/memo`, or send one copy of the manuscript to the following address: Centralized Manuscript Processing, MEMOIRS OF THE AMS, 201 Charles Street, Providence, RI 02904-2294 USA. If a paper copy is being forwarded to the AMS, indicate that it is for *Memoirs* and include the name of the corresponding author, contact information such as email address or mailing address, and the name of an appropriate Editor to review the paper (see the list of Editors below).

The paper must contain a *descriptive title* and an *abstract* that summarizes the article in language suitable for workers in the general field (algebra, analysis, etc.). The *descriptive title* should be short, but informative; useless or vague phrases such as "some remarks about" or "concerning" should be avoided. The *abstract* should be at least one complete sentence, and at most 300 words. Included with the footnotes to the paper should be the 2010 *Mathematics Subject Classification* representing the primary and secondary subjects of the article. The classifications are accessible from `www.ams.org/msc/`. The Mathematics Subject Classification footnote may be followed by a list of *key words and phrases* describing the subject matter of the article and taken from it. Journal abbreviations used in bibliographies are listed in the latest *Mathematical Reviews* annual index. The series abbreviations are also accessible from `www.ams.org/msnhtml/serials.pdf`. To help in preparing and verifying references, the AMS offers MR Lookup, a Reference Tool for Linking, at `www.ams.org/mrlookup/`.

Electronically prepared manuscripts. The AMS encourages electronically prepared manuscripts, with a strong preference for $\mathcal{A}\mathcal{M}\mathcal{S}$-LaTeX. To this end, the Society has prepared $\mathcal{A}\mathcal{M}\mathcal{S}$-LaTeX author packages for each AMS publication. Author packages include instructions for preparing electronic manuscripts, samples, and a style file that generates the particular design specifications of that publication series. Though $\mathcal{A}\mathcal{M}\mathcal{S}$-LaTeX is the highly preferred format of TeX, author packages are also available in $\mathcal{A}\mathcal{M}\mathcal{S}$-TeX.

Authors may retrieve an author package for *Memoirs of the AMS* from `www.ams.org/journals/memo/memoauthorpac.html`. The *AMS Author Handbook* is available in PDF format from the author package link. The author package can also be obtained free of charge by sending email to `tech-support@ams.org` or from the Publication Division, American Mathematical Society, 201 Charles St., Providence, RI 02904-2294, USA. When requesting an author package, please specify $\mathcal{A}_\mathcal{M}\mathcal{S}$-LaTeX or $\mathcal{A}_\mathcal{M}\mathcal{S}$-TeX and the publication in which your paper will appear. Please be sure to include your complete mailing address.

After acceptance. The source files for the final version of the electronic manuscript should be sent to the Providence office immediately after the paper has been accepted for publication. The author should also submit a PDF of the final version of the paper to the editor, who will forward a copy to the Providence office.

Accepted electronically prepared files can be submitted via the web at `www.ams.org/submit-book-journal/`, sent via FTP, or sent on CD to the Electronic Prepress Department, American Mathematical Society, 201 Charles Street, Providence, RI 02904-2294 USA. TeX source files and graphic files can be transferred over the Internet by FTP to the Internet node `ftp.ams.org` (130.44.1.100). When sending a manuscript electronically via CD, please be sure to include a message indicating that the paper is for the *Memoirs*.

Electronic graphics. Comprehensive instructions on preparing graphics are available at `www.ams.org/authors/journals.html`. A few of the major requirements are given here.

Submit files for graphics as EPS (Encapsulated PostScript) files. This includes graphics originated via a graphics application as well as scanned photographs or other computer-generated images. If this is not possible, TIFF files are acceptable as long as they can be opened in Adobe Photoshop or Illustrator.

Authors using graphics packages for the creation of electronic art should also avoid the use of any lines thinner than 0.5 points in width. Many graphics packages allow the user to specify a "hairline" for a very thin line. Hairlines often look acceptable when proofed on a typical laser printer. However, when produced on a high-resolution laser imagesetter, hairlines become nearly invisible and will be lost entirely in the final printing process.

Screens should be set to values between 15% and 85%. Screens which fall outside of this range are too light or too dark to print correctly. Variations of screens within a graphic should be no less than 10%.

Any graphics created in color will be rendered in grayscale for the printed version unless color printing is authorized by the Managing Editor and the Publisher. In general, color graphics will appear in color in the online version.

Inquiries. Any inquiries concerning a paper that has been accepted for publication should be sent to `memo-query@ams.org` or directly to the Electronic Prepress Department, American Mathematical Society, 201 Charles St., Providence, RI 02904-2294 USA.

Editors

This journal is designed particularly for long research papers, normally at least 80 pages in length, and groups of cognate papers in pure and applied mathematics. Papers intended for publication in the *Memoirs* should be addressed to one of the following editors. The AMS uses Centralized Manuscript Processing for initial submissions to AMS journals. Authors should follow instructions listed on the Initial Submission page found at www.ams.org/memo/memosubmit.html.

SELECTED PUBLISHED TITLES IN THIS SERIES

1323 **Shu Kawaguchi and Kazuhiko Yamaki,** Effective Faithful Tropicalizations Associated to Linear Systems on Curves, 2021

1322 **D. Bulacu and B. Torrecillas,** Galois and Cleft Monoidal Cowreaths. Applications, 2021

1321 **Christian Haase, Andreas Paffenholz, Lindsey C. Piechnik, and Francisco Santos,** Existence of Unimodular Triangulations–Positive Results, 2021

1320 **Th. Heidersdorf and R. Weissauer,** Cohomological Tensor Functors on Representations of the General Linear Supergroup, 2021

1319 **Abed Bounemoura and Jacques Féjoz,** Hamiltonian Perturbation Theory for Ultra-Differentiable Functions, 2021

1318 **Chao Wang, Zhifei Zhang, Weiren Zhao, and Yunrui Zheng,** Local Well-Posedness and Break-Down Criterion of the Incompressible Euler Equations with Free Boundary, 2021

1317 **Eric M. Rains and S. Ole Warnaar,** Bounded Littlewood Identities, 2021

1316 **Ulrich Bunke and David Gepner,** Differential Function Spectra, the Differential Becker-Gottlieb Transfer, and Applications to Differential Algebraic K-Theory, 2021

1315 **S. Grivaux, É. Matheron, and Q. Menet,** Linear Dynamical Systems on Hilbert Spaces: Typical Properties and Explicit Examples, 2021

1314 **Pierre Albin, Frédéric Rochon, and David Sher,** Resolvent, Heat Kernel, and Torsion under Degeneration to Fibered Cusps, 2021

1313 **Paul Godin,** The 2D Compressible Euler Equations in Bounded Impermeable Domains with Corners, 2021

1312 **Patrick Delorme, Pascale Harinck, and Yiannis Sakellaridis,** Paley-Wiener Theorems for a p-Adic Spherical Variety, 2021

1311 **Lyudmila Korobenko, Cristian Rios, Eric Sawyer, and Ruipeng Shen,** Local Boundedness, Maximum Principles, and Continuity of Solutions to Infinitely Degenerate Elliptic Equations with Rough Coefficients, 2021

1310 **Hiroshi Iritani, Todor Milanov, Yongbin Ruan, and Yefeng Shen,** Gromov-Witten Theory of Quotients of Fermat Calabi-Yau Varieties, 2021

1309 **Jérémie Chalopin, Victor Chepoi, Hiroshi Hirai, and Damian Osajda,** Weakly Modular Graphs and Nonpositive Curvature, 2020

1308 **Christopher L. Douglas, Christopher Schommer-Pries, and Noah Snyder,** Dualizable Tensor Categories, 2020

1307 **Adam R. Thomas,** The Irreducible Subgroups of Exceptional Algebraic Groups, 2020

1306 **Kazuyuki Hatada,** Hecke Operators and Systems of Eigenvalues on Siegel Cusp Forms, 2020

1305 **Bogdan Ion and Siddhartha Sahi,** Double Affine Hecke Algebras and Congruence Groups, 2020

1304 **Matthias Fischmann, Andreas Juhl, and Petr Somberg,** Conformal Symmetry Breaking Differential Operators on Differential Forms, 2020

1303 **Zhi Qi,** Theory of Fundamental Bessel Functions of High Rank, 2020

1302 **Paul M. N. Feehan and Manousos Maridakis,** Łojasiewicz-Simon Gradient Inequalities for Coupled Yang-Mills Energy Functionals, 2020

1301 **Joachim Krieger,** On Stability of Type II Blow Up for the Critical Nonlinear Wave Equation in $\mathbb{R}^{3+1}$, 2020

1300 **Camille Male,** Traffic Distributions and Independence: Permutation Invariant Random Matrices and the Three Notions of Independence, 2020